The Relay Testing Handbook

End-to-End Testing

The Relay Testing Handbook

End-to-End Testing

Chris Werstiuk

Professional Engineer
Journeyman Power System Electrician
Electrical Technologist

Valence Electrical Training Services

5005 S Kipling Pkwy Suite A7 PMB 395
Littleton, CO 80127-1375
https://RelayTraining.com

First printing in 2013.

The Relay Testing Handbook: End-to-End Testing

ISBN: 978-1-934348-15-4

Library of Congress Control Number: 2013932352

Published By: **Valence Electrical Training Services**
5005 S Kipling Pkwy Suite A7 PMB 395, Littleton, CO 80127-1375, U.S.A. | (303) 250-8257 |

store@relaytraining.com

Distributed by: **https://relaytraining.com**

Edited by: **One-on-One Book Production, West Hills, CA**

Cover Art: **© James Steidl. Image from BigStockPhoto.com**

Interior Design and Layout: **Adina Cucicov, Flamingo Designs**

Printed in the United States of America.

Author's Note

Traditional protective relay books are written by engineers for engineers to use when modeling the electrical system or creating relay settings, and they often have very little practical use for the test technician in the field. *The Relay Testing Handbook* is a practical resource written by a relay tester for relay testers; it is a comprehensive series of practical instructional manuals that provides the knowledge necessary to test most modern protective relays. The complete handbook combines basic electrical fundamentals, detailed descriptions of protective elements, and generic test plans with examples of real-world applications, enabling you to confidently handle nearly any relay testing situation. Practical examples include a wide variety of relay manufacturers and models to demonstrate that you can apply the same basic fundamentals to most relay testing scenarios.

This book provides an overview of end-to-end testing and answers the most common questions a relay tester should ask before performing their first end-to-end test. A basic introduction of this test technique is followed by a step-by-step procedure for performing a successful end-to-end test. This package also includes an overview of the most common communication-assisted protection schemes to help the reader understand how these schemes operate.

Thank you for supporting this major undertaking. I hope you find this and all other installments of *The Relay Testing Handbook* series to be a useful resource. This project is ongoing and we are constantly seeking to make improvements. Our publishing model allows us to quickly correct errors or omissions and implement suggestions. Please contact us at store@relaytraining.com to report a problem. If we implement your suggestion, we'll send you an updated copy and a prize. You can also go to **https://relaytraining.com/updates** to see what's changed since the *The Relay Testing Handbook* was released in 2012.

Acknowledgments

This book would not be possible without support from these fine people:

DAVID MAGNAN, PROJECT MANAGER
PCA Valence Engineering Technologies Ltd.
www.pcavalence.com

KEN GIBBS, C.E.T.
PCA Valence Engineering Technologies Ltd.

LES WARNER C.E.T.
PCA Valence Engineering Technologies Ltd.

ROGER GRYLLS, CET
Magna IV Engineering
Superior Client Service. Practical Solutions
www.magnaiv.com

Eric Cameron, B.E.Sc.
Ainsworth Power Services
www.ainsworth.com

Philip B. Baker
Electrical Technician

David Snyder
Hydropower Test and Evaluation
Bonneville Lock and Dam
www.nwp.usace.army.mil/op/b/

Todd J. Patterson
Engineering Technologist
Protection and Control Training Coordinator
American Electric Power

Table of Contents

Table of Figures

Chapter 1

Introduction to End-to-End Testing

End-to-end testing is considered to be the ultimate test for protection schemes that have two or more protective relays communicating trip and blocking information with each other in order to provide more accurate fault detection. End-to-end testing can prove relay protection schemes before placing them into service by:

- temporarily installing a protective relay test-set at each location to simulate currents and voltages,
- synchronizing the test-sets to the same time with high accuracy,
- applying a simulated fault using unique current and voltage values created for each test location,
- and evaluating the relays' responses.

Modern relay test equipment has helped to make this a more popular test technique for testing advanced protection schemes, especially as the National Electrical Reliability Council (NERC) and other regulatory agency standards are becoming more stringent. One excerpt from the NERC requirements includes the following text that almost requires end-to-end testing to be performed on every new installation: "*At installation, the acceptance test should be performed on the complete relay scheme in addition to each individual component so that the adequacy of the scheme is verified.*"

This book will introduce the theoretical and practical aspects of end-to-end testing that every relay tester should know before attempting their first end-to-end test. As we usually stress in the *The Relay Testing Handbook* series, it is important to understand what you are testing and why, so we will start with some basic line protection theory.

1. How to Protect Transmission Lines?

Unlike distribution feeders located at industrial sites or end-user locations, most transmission lines are connected with sources at both ends of the line, which can cause power swings in either direction that can be very dramatic. The ideal protection scheme for a transmission line will ignore load swings or external faults, and only operate if a fault occurs on the transmission line.

The ideal solution would measure the current entering and leaving the transmission line, and trip if the two sides did not match. Differential relays sound like the perfect solution; but remember that the transmission line can be hundreds of miles long and you need to communicate information about each phase back and forth between the two sides within microseconds. You would also need to:

- install a direct communication path,
- generate a signal that will not degrade over long distances,
- generate analog information (current magnitudes and angles) across that signal,
- use a device that can convert the signal into meaningful data,
- and account for the time the signal spent traveling between the two devices.

Modern line differential relays use fiber optic communication and microprocessors to overcome these obstacles and are probably the best protection available today. However, those technologies were not available in the early days of electrical protection. Creative engineers developed some solutions including the phase charge comparison relays discussed later in this book, but these solutions still required additional infrastructure (special relays and communication channels), additional cost, and additional engineering not available to all utilities.

2. What are Impedance Relays?

Early protection engineers realized that they could provide pretty good transmission protection by measuring or calculating the transmission line's impedance, and then monitoring the power system impedance at each end of the transmission line. The design engineers reasoned that if the measured impedance was less than the calculated impedance, then there must be a fault on the transmission line, and the relay should trip instantaneously. The first impedance relays were constructed with two coils that would "calculate" the impedance. A voltage coil would pull the two contacts apart and a current coil would attempt to pull the contacts together. Both coils had taps that varied the amount of force each coil applied so they could vary the impedance pickup. The relay picks up when the current coil force is stronger than the voltage coil because any fault will cause the measured voltage to decrease with a corresponding increase of current. They now had a way to apply Ohm's Law with magnetism and mechanisms that created a circular operating characteristic called the mho circle or characteristic.

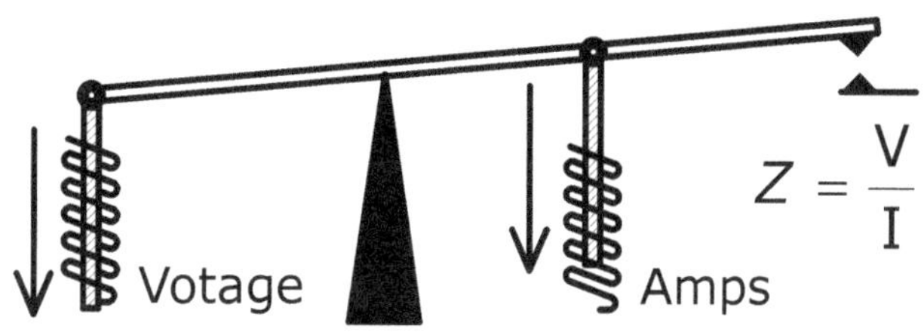

Figure 1-1: Electromechanical Impedance Protection

Impedance protection evolved to provide directional control; and most of today's relays typically use the circle or mho characteristic as shown in Figure 1-2.

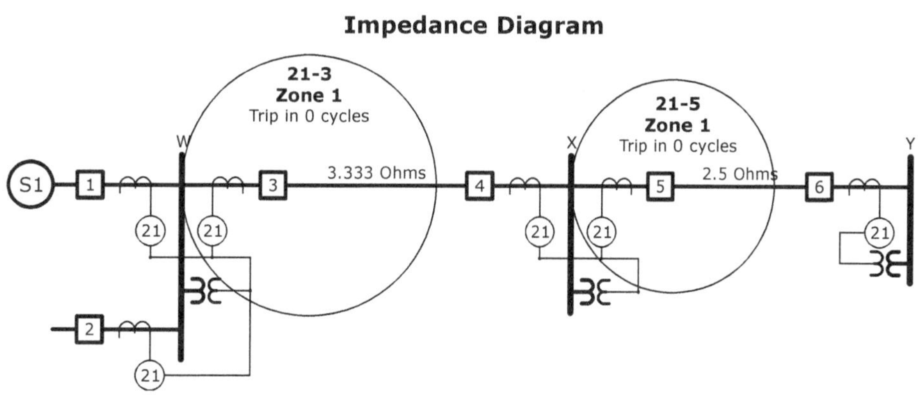

Figure 1-2: Zone-1 Impedance Protection

You should notice that the 21-3 relay's mho characteristic in Figure 1-2 does not protect the entire transmission line and is usually defined as Zone-1 impedance protection. Most impedance relays are set to instantaneously trip when the measured impedance is between 75-90% of the transmission line impedance to compensate for calculation errors and, more importantly, the 10% accuracy class typical for most protection class CTs. A second, larger mho characteristic is usually applied as Zone-2 with a 15-30 cycle time delay to provide 100% line protection, and backup protection for adjacent lines by allowing the other Zone-1 relays a chance to operate. If the other relays cannot isolate the fault, the Zone-2 protection on the adjacent line will operate.

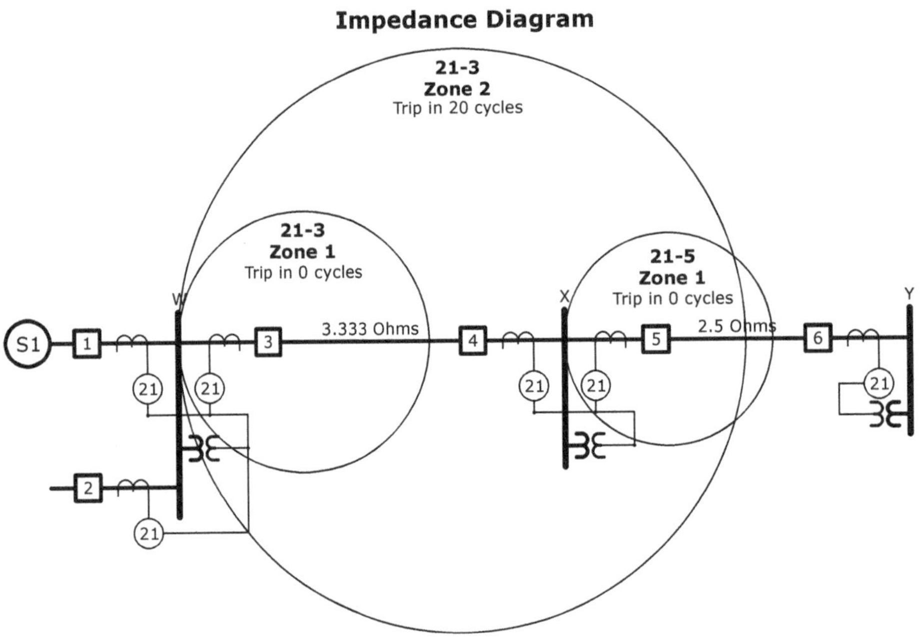

Figure 1-3: Zone-1 and Zone-2 Impedance Protection

Traditional line protection schemes (Figure 1-4) use an impedance relay on either side of the transmission line to constantly monitor voltage and current to calculate impedance in real-time. If a fault occurs in the overlap area between the Zone-1s of both relays, both relays will trip instantaneously and isolate the fault from the system. A perfect solution! However, if the fault occurs in the first 10-25% percent of a transmission line, the relay closest to the fault will operate instantaneously but the other relay will wait for the specified Zone-2 time delay. 15-30 cycles may not seem like a long time, but it can be forever in electrical terms because a lot of damage and system disruption can occur in that short time frame.

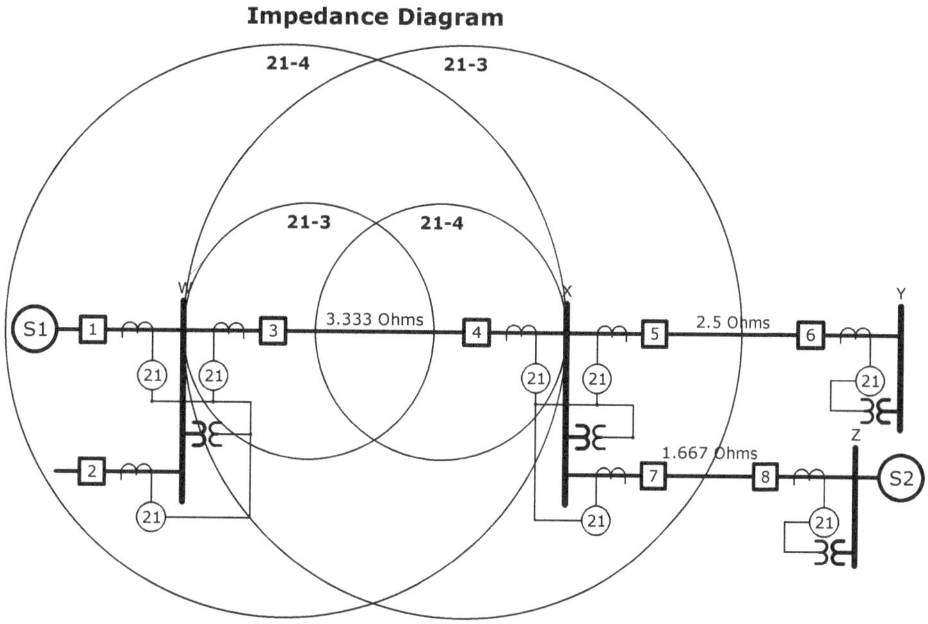

Figure 1-4: Complete Zone-1 and Zone-2 Impedance Protection

Every mis-operation costs utilities labor and lost revenue, so 75-90% protection is often not good enough, even though both sides overlap to provide a pretty good protection scheme. Modern protection schemes apply a communication channel between the relays on either side of a transmission line, which allows the relays to communicate with each other. The simplest scheme (not really, but easiest to understand) applies the differential protection principles discussed earlier; each relay measures the current entering or leaving the transmission line and shares its local current with the remote relay bidirectionally. If the current-in does not match the current-out, both relays will trip. These relays often have back-up schemes using traditional impedance protection that is only active if the relays are unable to communicate.

The other communication schemes have many names (DUTT, PUTT, POTT, DCUB, etc.) and operating parameters, but they all perform the same basic function. Both relays monitor the real-time impedance and current direction. If both relays agree the fault is between them, both relays will trip as soon as possible depending on the communication medium and scheme. These methods allowed utilities to provide 100% protection of a transmission line by transmitting digital signals (On/Off) between substations, which is a far simpler and more cost-effective task compared to transmitting analog signals (magnitudes and vectors).

3. What is End-to-End Testing?

End-to-end testing evaluates the entire protective relay scheme as a whole using two or more test-sets at multiple locations to simultaneously simulate a fault at every end of a transmission line. This test technique previously required specialized knowledge and equipment to perform, but modern test-sets make it a relatively simple task. Figure 1-5 represents an overview of the equipment and personnel required for a typical end-to-end test using a simple transmission line with two ends or, as they're sometimes called, nodes. It is possible to have a system with three or more nodes which simply adds another location to the test plan.

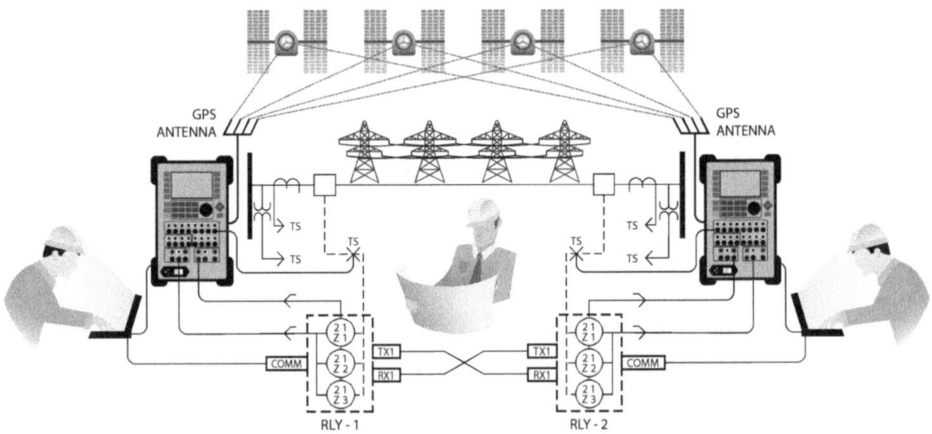

Figure 1-5: End-to-End Testing Summary

The following components are necessary for the relay tester to perform a successful end-to-end test:

1. A relay test-set for each location with a minimum of:
    - three voltage channels,
    - three current channels,
    - at least one programmable output to simulate breaker status or other external signals,
    - at least one programmable input to detect trip or breaker status signals,
    - an internal GPS clock (some test-sets allow for other time signals such as IRIG), or an external GPS clock with an output signal and an additional test-set start input,
    - and waveform playback capabilities or a Fault-state/state-simulator with at least 3 programmable states.
2. Some test-sets require a computer to control the test-set playback or state functions.
3. A computer and software to download and display event records obtained from the relay after each test.

4. At least one relay tester at each location with some form of communication between them such as telephone or over-network communication. It is possible, but not recommended, for one person to perform all tests if the relay, test-sets, and communication systems have all been configured properly.
5. The engineer(s) who created the settings and test plans should be on call to correct any errors discovered during the tests.
6. A setting file, waveform, or detailed description of the specific test scenarios.
7. An understanding of the relay protection scheme and what the expected result for each test should be.

The relay testers at each end of the line perform the following steps during an end-to-end test:

1. Obtain test cases from the engineer and review them to find obvious errors and determine what the expected results should be.
2. Set up the GPS Antenna and apply GPS time as the test-set reference (or use other reference such as IRIG, if required.)
3. Isolate the equipment under test.
4. Connect the appropriate input and output connections.
5. Connect the test-set to replace Current Transformer (CT)/Potential Transformer (PT) connections.
6. Communicate with remote testers and apply a meter test on all sides. Verify correct results.
7. Communicate with remote sides and determine which test plan will be used for the test.
8. Load the appropriate test cases into all test-sets.
9. Place all circuit breakers in the correct positions, or ensure circuit breaker contacts are properly simulated by the test-set.
10. Communicate the start time with all sides and initiate the test.
11. Review targets for correct operation and download all event records. Review event records for correct operation, if required.
12. Repeat from Step 7-11 for all test cases.
13. Restore the equipment to service.

4. Why Do We Perform End-to-End Testing?

The most effective transmission line protection using modern technology is achieved by installing protective relays at each end of the line that constantly trade information about the power system through a communication channel. Any disturbance is communicated to the other relays, which will cause the protection to operate more quickly depending on the protection scheme used and the apparent location of the fault. These protection schemes, when applied correctly, can make the transmission line protection more reliable and more selective than is possible with a single relay or a series of relays that cannot communicate.

While it is possible to test each of the individual components separately, many problems can only be detected when the entire scheme is tested as a whole. It is possible to test one side at a time, which can give the tester a reasonable sense that the scheme will operate successfully on proven relay settings. However, many problems with communication-assisted protection occur when the fault changes directions, or by incorrectly defined communication delays that are inherent in the system. These problems can only be detected by properly applied end-to-end testing or a review of an incorrect relay operations after a fault.

End-to-end testing was considered a daunting task a decade ago, but advances in relay testing technology and personal computers have reduced the complexity to a couple of extra steps for a reasonably experienced relay tester.

5. How Does it Work?

Most system disturbances occur within one millisecond and modern protective relays must be able to detect faults within this time frame to be effective. Practical experience has shown that two test-sets must start within 10 microseconds of each other to provide reliable results. This causes a problem for multiple relay testers at multiple locations because it is nearly impossible for two relay testers at two different locations to press start within 10 microseconds. The remote relay testers could use the power system itself to synchronize the two test-sets, but they could introduce up to 1 ms or 22° error to the test which is outside the ten microsecond tolerance required for consistent results. It would be difficult to determine whether the protective system operated because of a problem, or the difference between start times.

End-to-end testing became a viable test technique for everyone when the U.S. Armed Forces allowed non-military access to the time signals sent by a system of satellites that become the Global Positioning System (GPS). This system operates by obtaining the time signal and general location of at least four satellites and comparing the differences between time and distance to determine the antenna's location within a few meters. Time can be synchronized within one microsecond anywhere that four satellites are available. Most modern test equipment can specify synchronization within 2 microseconds, which is within the maximum allowable time delay by a factor of five.

Once two test-sets have synchronized time sources, at least two Fault states are applied with information from a fault simulator program. The first Fault state injects the normal currents, voltages, and phase angles for the transmission line to create a Prefault or normal condition. After a pre-determined time, both test-sets switch to a Fault state that simulate the voltages, currents, and phase angles of a system fault. It is important to note that every side of the transmission line will have different values depending on the location of the fault and the power sources around the transmission line. It is vital that the correct fault simulations be applied to the correct relays. These Fault states could be combined into one file, typically COMTRADE (I.E.E.E. standard IEEE C37.111 for waveform files), and played back into the relay; or created using fault information and manually entered into the fault simulations. If the fault has been properly simulated at all locations simultaneously, the protective relays should operate as if a fault occurred on the system. The results should be evaluated to ensure the protective relays and communication equipment are functioning correctly as a unit.

It is also important to note that different test-set manufacturers, different test-set models, and different test-set firmware may be synchronized to the same time source but may not start outputting the test at the same moment due to internal time delays and/or external I/O time delays, if used. Always consult with the relay test-set manufacturers if two different models of test-sets will be used for end-to-end testing to determine if a correction factor must be applied. Different models from the same manufacturer can produce different starting times, and the correction factors should be verified at the same location, if possible, before performing any remote testing.

6. Where Should I Perform End-to-End Testing?

End-to-end testing should be performed whenever it would be beneficial to test an entire protection scheme in real-time to make sure all the equipment will operate correctly when required. This test technique need not be limited to transmission lines and can be applied any time you wish to test coordination between different devices. For example, a number of test-sets can be connected to the protective relays for all feeders in a system. A fault simulation can be played into all relays simultaneously to ensure that complex blocking schemes work as intended.

7. When Should I Perform End-to-End Testing?

End-to-end testing appears to be mandated by the NERC requirement quoted earlier, and all new installations with remote communication between relays should be tested via end-to-end testing. This test technique can also be a useful and effective maintenance test if end-to-end testing was performed during the relays' commissioning. There can be no more effective way of ensuring the entire protection scheme than re-playing the same number of tests into the protection system and observing the same results. Performed correctly, using this test technique for maintenance tests can be more efficient as well.

Chapter 2

Detailed End-to-End Testing Procedures

This section will provide more detailed information about each step of the end-to-end testing procedure described as:

1. Obtain and Review All Test Cases
2. Set up the GPS Antenna
3. Isolate the Equipment Under Test
4. Relay Input and Output Connections
5. Connect Test Equipment To Replace CTs/PTs
6. Apply a Meter Test
7. Apply the Test Plan
8. Evaluate the Results
9. Repeat Steps 7-8 for All Tests
10. Return Protection System to Service
11. Prepare the Report

1. Obtain and Review All Test Cases

End-to-end testing is performed by simultaneously applying different test cases to multiple relays that will simulate faults at various locations along and around the transmission line using a mix of the most common fault types (Phase to Neutral, Phase to Phase, and Three-Phase). Additional test cases are performed outside the protected zone to ensure that the relay will not trip.

The traditional settings for a distance relay are displayed with a 10Ω transmission line in Figure 2-1 where:

- Zone-1 = 80% of the transmission line with no intentional delay
- Zone-2 = 125% of the transmission line with a fixed delay around 20 cycles
- Zone-3 = Site-specific percentage in the reverse direction

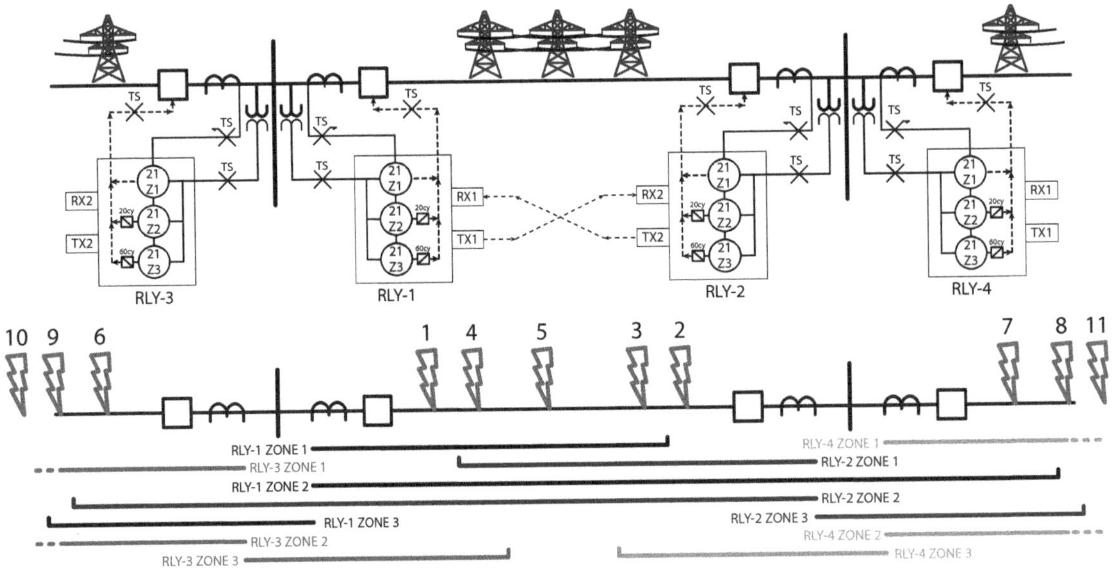

Figure 2-1: Typical Distance Protection Settings

A typical series of tests will verify the relays' operation at key locations along the transmission line, usually slightly above and below the pickup settings for each zone. For example, Tests 1 and 4 would be performed at 15% (1.5Ω) and 25% (2.5Ω) from RLY-1 to test the Zone-1 protection boundaries. Tests 2 and 3 verifies RLY-2's Zone-1 protection boundaries by applying faults at 85% and 75% from RLY-1, or 1.5Ω and 2.5Ω from RLY-2. Another four tests (6, 7, 8, and 9) would be performed to test the Zone-2 protection of both relays at 120% and 130% from relays RLY-1 and RLY-2. Another two to four tests are performed to test Zone-3, if Zone-3 is enabled. A final test can be performed to ensure that the relays will not operate due to a sudden phase reversal when a fault occurs on one of two parallel lines.

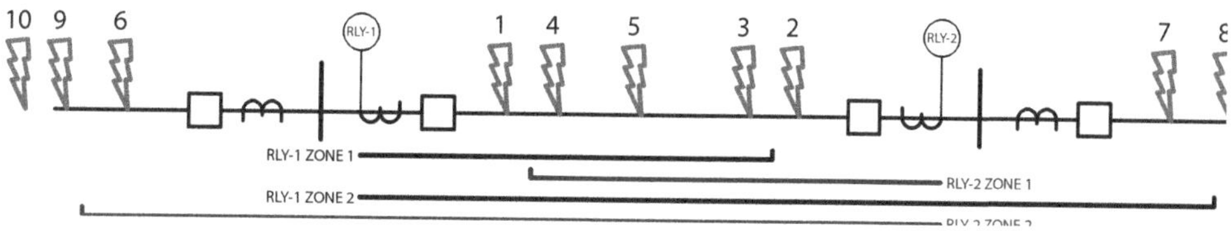

Figure 2-2: Typical End-to-End Tests

It is possible to create test plans by manually calculating all of these points on a radial (only one source) transmission line knowing the line length and settings. However, these manual calculations will not include the source impedance, which significantly affects the Prefault and Fault values. Proper fault calculations become very complicated when the transmission line has more than one source available, which makes test case creation beyond the average relay tester. You should always obtain test cases from the design engineer when performing end-to-end tests on any distance protection scheme. It should be relatively easy for the design engineer to choose the test case parameters in their modelling software, and export the results as a waveform or data file. The test cases should be submitted to the testing team with a description of the expected results for each test, another reason why this information should be supplied by the design engineer so you can test the site-specific settings instead of the relay's programming.

Relay-testers normally see a magnitude, angle, and time delay for each channel, and the test-set converts those commands into a waveform to be generated into the relay. Waveform files specify every point on the waveform to be generated, and they can be the simplest way (when everything is working correctly) to perform end-to-end testing. Waveform files can also provide complex simulations of system distortions (transients, DC offset, or CVT distortions) that typically occur during a real fault. Waveform data is created when the design engineer exports the data into COMTRADE format for each test case, and sends the files to the relay tester. The relay testers open the file in their respective test-sets, check to ensure the correct channels are used, and run the tests. If the relays respond correctly, the relay testers save the data and move on to the next test. However, it can be difficult for two relay testers at two different locations to troubleshoot problems in the test plan itself because they cannot compare the two waveforms side-by-side to find any gross errors.

Supplying test cases as data can be tedious for the design engineer and the relay tester, depending on the system modeling software used and the intended test-set. Ideally, the design engineer exports the data as a file that defines the magnitude, angle, and time to be injected into each channel, and that file is imported into the test-set without difficulty. This ideal situation is often not the case and some intermediary software may be required for the conversion process. The information could also be manually entered into the test-set software. This method provides better documentation of the tests and allows the relay testers at different locations to quickly compare test cases to find gross errors. However, this method can be more time consuming and prone to simple conversion errors.

An example test case is shown below using both methods.

RLY-1	Load from Bus to Line (SOURCE BUS for Prefault)													
	Aspen Output Pre		Aspen Output Fault 1		Aspen Output Fault		Aspen Output Post		PRE FAULT Secondary Test Set		FAULT 1 Secondary Test Set		Post Fault Secondary Test Set	
	Values	Angle	Values	Angle	Values	Angle	Values	Angle						
VA	132.79	0.00	110.00	-1.00			132.79	0.00	**66.40**	**0.00**	**55.50**	**-1.00**	**66.40**	**0.00**
VB	132.79	-120.00	110.00	-121.00			132.79	-120.00	**66.40**	**-120.00**	**55.50**	**-121.00**	**66.40**	**-120.00**
VC	132.79	120.00	110.00	119.00			132.79	120.00	**66.40**	**120.00**	**55.50**	**119.00**	**66.40**	**120.00**
IA	200.00	-10.00	4452.00	-81.00			0.00	0.00	**0.50**	**-10.00**	**11.13**	**-81.00**	**0.00**	**0.00**
IB	200.00	-130.00	4452.00	159.00			0.00	0.00	**0.50**	**-130.00**	**11.13**	**159.00**	**0.00**	**0.00**
IC	200.00	110.00	4452.00	39.00			0.00	0.00	**0.50**	**110.00**	**11.13**	**39.00**	**0.00**	**0.00**
Time (cy)	30.00		65.00				30.00		30.00		65.00		30.00	

RLY-2	Load from Line to Bus (LOAD BUS for Prefault)													
	Aspen Output Pre		Aspen Output Fault 1		Aspen Output Fault		Aspen Output Post		PRE FAULT Secondary Test Set		FAULT 1 Secondary Test Set		Post Fault Secondary Test Set	
	Values	Angle	Values	Angle	Values	Angle	Values	Angle						
VA	132.79	0.00	119.17	-1.00			132.79	0.00	**66.40**	**0.00**	**59.60**	**-1.00**	**66.40**	**0.00**
VB	132.79	-120.00	119.17	-121.00			132.79	-120.00	**66.40**	**-120.00**	**59.60**	**-121.00**	**66.40**	**-120.00**
VC	132.79	120.00	119.17	119.00			132.79	120.00	**66.40**	**120.00**	**59.60**	**119.00**	**66.40**	**120.00**
IA	200.00	170.00	4784.00	-81.00			0.00	0.00	**0.33**	**170.00**	**7.97**	**-81.00**	**0.00**	**0.00**
IB	200.00	50.00	4784.00	159.00			0.00	0.00	**0.33**	**50.00**	**7.97**	**159.00**	**0.00**	**0.00**
IC	200.00	-70.00	4784.00	39.00			0.00	0.00	**0.33**	**-70.00**	**7.97**	**39.00**	**0.00**	**0.00**
Time (cy)	30.00		65.00				30.00		30.00		65.00		30.00	

Figure 2-3: Example Test Plan Using Raw Data

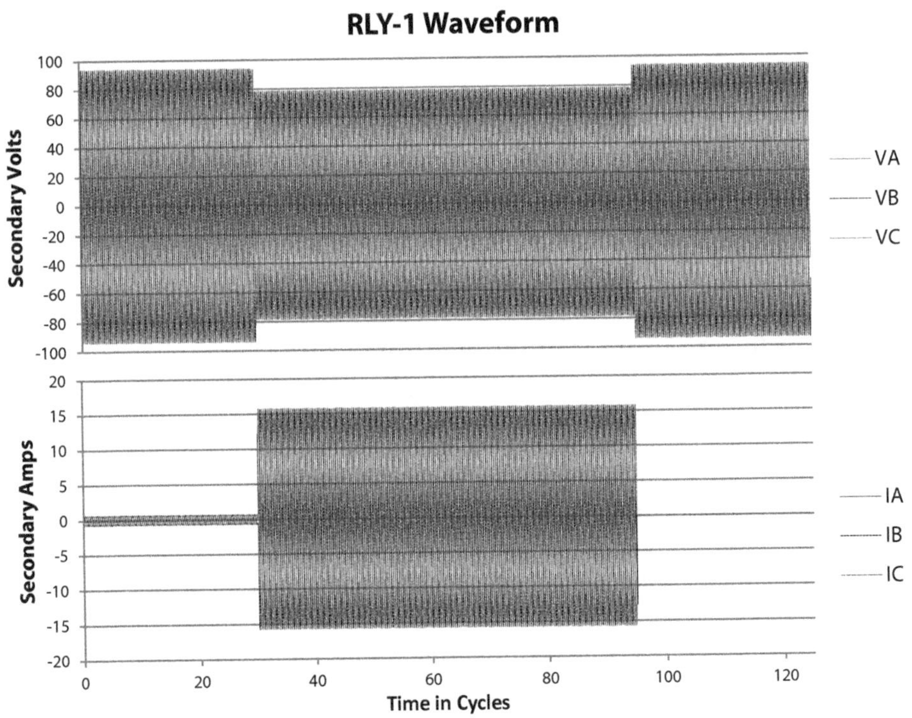

Figure 2-4: RLY-1 Test Case as Waveform

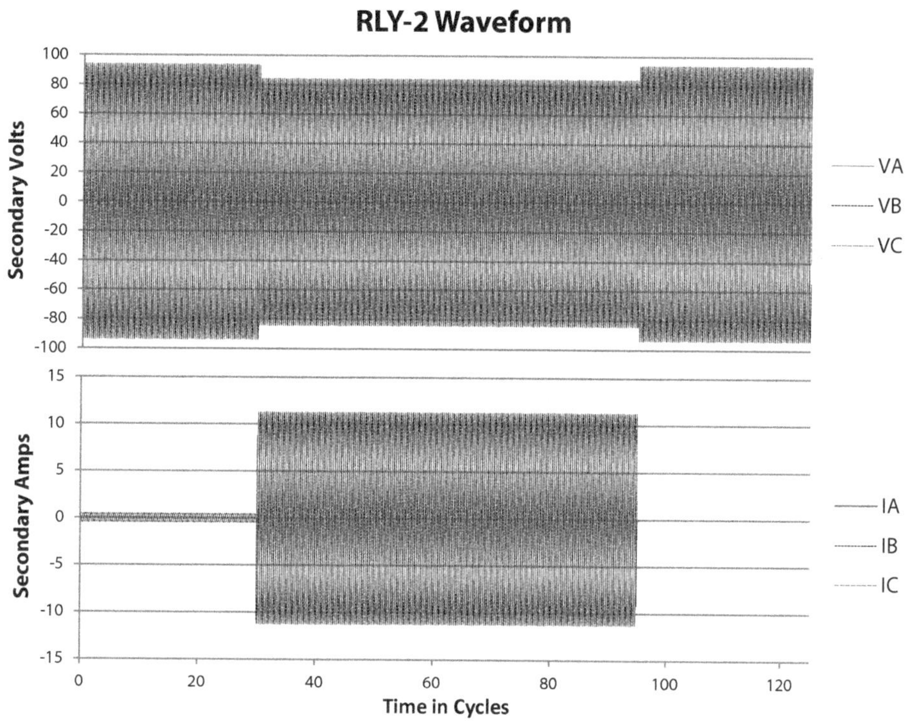

Figure 2-5: RLY-2 Test Case as Waveform

The following steps should be performed when reviewing any files supplied by the design engineer:

A) Are the Files Labeled Correctly and Do They Include Expected Results?

Review the file names and make sure they make sense for your application. The file name should include the following:

1. Which relay the file should be applied to (RLY-1)
2. What test case number (RLY-1_Test1)

Many file names also have additional information such as:

1. Fault type (RLY-1_Test1_A-N)
2. Fault location (RLY-1_Test1_A-N_5%fromRLY-1)
3. Expected results (RLY-1_Test1_A-N_5%fromRLY-1_Zone1 or RLY-1_Test9_3P_95%fromRLY-1_Comm=6cycles)

There should also be a summary of the tests along with the files describing exactly what should happen when each test is applied with, at a minimum, the information if the following example.

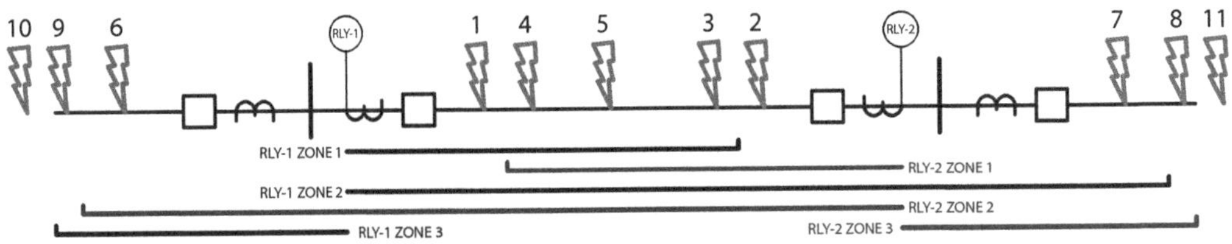

Figure 2-6: End-to-End Test Example

Test Case #1

- Three-Phase, In-section, 15% from RLY-1
- RLY-1 Results = 15 Miles, <3cycle Trip, Zone-1 & A-Phase & B-Phase & C-Phase Target, Key Permission
- RLY-2 Results = 85 Miles, <6cycle Trip, Pilot Trip & Zone-2 & A-Phase & B-Phase & C-Phase Target, Key Permission

Test Case #2

- B-Phase to C-Phase, In-section, 85% from RLY-1
- RLY-1 Results = 85 Miles, <6cycle Trip, Pilot Trip & Zone-2 & B-Phase & C-Phase Target, Key Permission
- RLY-2 Results = 15 Miles, <3cycle Trip, Zone-1 & B-Phase & C-Phase Target, Key Permission

Test Case #3

- A-Phase to Ground, In-section, 75% from RLY-1
- RLY-1 Results = 75 Miles, <3cycle Trip, Zone-1 & A-Phase Target, Key Permission
- RLY-2 Results = 25 Miles, <3cycle Trip, Zone-1 & A-Phase Target, Key Permission

Test Case #4

- B-Phase to Ground, In-section, 25% from RLY-1
- RLY-1 Results = 25 Miles, <3cycle Trip, Zone-1 & B-Phase Target, Key Permission
- RLY-2 Results = 75 Miles, <3cycle Trip, Zone-1 & B-Phase Target, Key Permission

Test Case #5

- Three-Phase, In-section, 50% from RLY-1
- RLY-1 Results = 50 Miles, <3cycle Trip, Zone-1 & A-Phase & B-Phase & C-Phase Target, Key Permission
- RLY-2 Results = 50 Miles, <3cycle Trip, Zone-1 & A-Phase & B-Phase & C-Phase Target, Key Permission

Test Case #6

- C-Phase to Ground, Out-of-section, 120% from RLY-2
- RLY-1 Results = Key Block and No Operation
- RLY-2 Results = 120 Miles, <23cycle Trip, Zone-2 & C-Phase Target, Key Permission

Test Case #7

- C-Phase to A-Phase, Out-of-section, 120% from RLY-1
- RLY-1 Results = 120 Miles, <23cycle Trip, Zone-2 & C-Phase & A-Phase Target, Key Permission
- RLY-2 Results = Key Block and No Operation

Test Case #8

- Three-Phase, Out-of-section, 130% from RLY-1
- RLY-1 Results = No Operation
- RLY-2 Results = Key Block and No Operation

Test Case #9

- A-Phase to B-Phase, Out-of-section, 130% from RLY-2
- RLY-1 Results = Key Block and No Operation
- RLY-2 Results = No Operation

Test Case #10

- A-Phase to Ground, Out-of-section, 150% from RLY-2
- RLY-1 Results = No Operation
- RLY-2 Results = No Operation

Test Case #11

- B-Phase to C-Phase, Out-of-section, 150% from RLY-1
- RLY-1 Results = No Operation
- RLY-2 Results = No Operation

Test Case #12

- Test Case #7 on C-Phase to A-Phase for 10 cycles then Test Case #6 C-Phase to A-Phase for 10 Cycles
- RLY-1 Results = Key Permission then Key Block, No Operation
- RLY-2 Results = Key Block then Key Permission, No Operation

Test Case #13

- Test Case #6 on A-Phase to Ground for 10 cycles then Test Case #7 on A-Phase to Ground for 10 Cycles
- RLY-1 Results = Key Block then Key Permission, No Operation
- RLY-2 Results = Key Permission then Key Block, No Operation

Or a summary table could be provided like the following:

CASE #	RELAY	PHASE	SECTION	LOCATION (%)	LOCATION (MILES)	TRIP (CYCLES)	TARGET	FUNCTION
1	RLY-1	3P	In	15%	15	<3	Z1, ABC	Key Perm
	RLY-2	3P	In	85%	85	<6	Z2, ABC, Comm	Key Perm
2	RLY-1	B-C	In	85%	85	<6	Z2, BC, Comm	Key Perm
	RLY-2	B-C	In	15%	15	<3	Z1, BC	Key Perm
3	RLY-1	A-N	In	75%	75	<3	Z1, AG	Key Perm
	RLY-2	A-N	In	25%	25	<3	Z1, AG	Key Perm
4	RLY-1	B-N	In	25%	25	<3	Z1, BG	Key Perm
	RLY-2	B-N	In	75%	75	<3	Z1, BG	Key Perm
5	RLY-1	3P	In	50%	50	<3	Z1, ABC	Key Perm
	RLY-2	3P	In	50%	50	<3	Z1, ABC	Key Perm
6	RLY-1	C-N	Out	-20%	N/A	N/A	N/A	Key Block
	RLY-2	C-N	Out	120%	120	<23	Z2, CN	Key Perm
7	RLY-1	C-A	Out	120%	120	<23	Z2, CA	Key Perm
	RLY-2	C-A	Out	-20%	N/A	N/A	N/A	Key Block
8	RLY-1	3P	Out	130%	N/A	N/A	N/A	N/A
	RLY-2	3P	Out	-30%	N/A	N/A	N/A	Key Block
9	RLY-1	A-B	Out	-30%	N/A	N/A	N/A	Key Block
	RLY-2	A-B	Out	130%	N/A	N/A	N/A	N/A
10	RLY-1	AG	Out	-50%	N/A	N/A	N/A	N/A
	RLY-2	AG	Out	150%	N/A	N/A	N/A	N/A
11	RLY-1	BC	Out	150%	N/A	N/A	N/A	N/A
	RLY-2	BC	Out	-50%	N/A	N/A	N/A	N/A
12	RLY-1	CA	Reversal	120 / -20	N/A	N/A	N/A	Key Perm / Key Block
	RLY-2	CA	Reversal	-20 / 120	N/A	N/A	N/A	Key Block / Key Perm
13	RLY-1	AG	Reversal	-20 / 120	N/A	N/A	N/A	Key Block / Key Perm
	RLY-2	AG	Reversal	120 / -20	N/A	N/A	N/A	Key Perm / Key Block

Figure 2-7: Example Test Plan Summary

B) Do the Files Contain Simple Mistakes?

This step may seem like overkill, but remember that on-site troubleshooting is much more difficult because you usually can't see the other side's information when testing. Spending a few minutes on this step will save you hours on test day! You should review each test with both test cases side by side on a split screen and look for the following common mistakes that are usually easy to spot:

i) Are Prefault, Fault1, and Fault 2, etc. Parameters Correct?

End-to-end tests can be performed with different operating scenarios and it is important that the functions are consistent. Some of the most common scenarios include:

Normal Test

- Inject a Prefault state (normal operating conditions) for a preset amount of time,
- Automatically switch to a Fault state for a preset amount of time,
- Automatically switch to Postfault state for a preset amount of time,
- Automatically stop.

Fault Simulation

- Inject a Prefault state for a preset amount of time,
- Automatically switch to a Fault state until the relay trips,
- Automatically stop.

Fault Simulation with Postfault

- Inject a Prefault state for a preset amount of time,
- Automatically switch to a Fault state until the relay trips,
- Automatically switch to a Postfault state for a preset amount of time,
- Automatically stop.

Fault Simulation with Breaker Timing Simulation

- Inject a Prefault state for a preset amount of time,
- Automatically switch to a Fault state until the relay trips,
- Continue injecting the Fault for a preset amount of time to simulate the breaker opening time,
- Automatically stop.

Fault Simulation with Breaker Timing Simulation with Postfault

- Inject a Prefault state for a preset amount of time,
- Automatically switch to a Fault state until the relay trips,
- Continue injecting the Fault for a preset amount of time to simulate the breaker opening time,
- Automatically switch to Postfault state for a preset amount of time,
- Automatically stop.

ii) Are Prefault, Fault1, and Fault 2, etc. Times Correct?

Simultaneously review all of the test simulations for each test case using your test-set's native software on a split screen to make sure you are looking at the files that will be generated and not what should be generated. Make sure that all time settings are identical. This is particularly important for Prefault states. Incorrectly applied Prefault times are particularly bad when testing line differential relays because one side can transition to the Fault state while the other side continues to generate a Prefault. This is the definition of a differential fault, so the relay(s) will trip. If you are not paying attention, you might think that the relays operated correctly when, in fact, there was a problem with the test plan.

This is an easy step using test data plans because the times will be displayed as numbers that you can read as shown in Figure 2-8:

RLY-1	Test Case #1							
	PREFAULT		FAULT 1		FAULT 2		Post Fault	
VA	66.4	0.0	67.3	0.0	67.3	0.0	66.4	0.0
VB	66.4	-120.0	33.7	180.0	52.5	-130.0	66.4	-120.0
VC	66.4	120.0	33.7	180.0	52.3	129.0	66.4	120.0
IA	0.50	-10.0	0.06	-85.0	0.06	-85.0	0.0	-10.0
IB	0.50	-130.0	4.65	9.0	3.66	-171.0	0.0	-130.0
IC	0.50	110.0	4.65	-172.0	3.66	10.0	0.0	110.0
Time (cy)	60.00		10.00		10.00		60.00	

RLY-2	Test Case #2							
	PREFAULT		FAULT 1		FAULT 2		Post Fault	
VA	66.4	0.0	66.7	0.0	66.7	0.0	66.4	0.0
VB	66.4	-120.0	57.5	-126.0	34.1	-173.0	66.4	-120.0
VC	66.4	120.0	56.7	125.0	33.1	173.0	66.4	120.0
IA	0.33	170.0	0.04	95.0	0.04	95.0	0.0	170.0
IB	0.33	50.0	3.10	-171.0	2.44	9.0	0.0	50.0
IC	0.33	-70.0	3.10	8.0	2.44	-170.0	0.0	-70.0
Time (cy)	60.00		10.00		10.00		60.00	

Figure 2-8: Example Test Plan Comparison

This step can be more complicated when looking at waveform files because your software may not allow you to view a time scale, which means you will be reduced to counting cycles. You may be able to overlap channels for both test plans on one graph, or make sure that the waveform scales are identical on both displays to make sure the times are consistent between test cases.

iii) Are Prefault State Voltages and Currents Correct?

You should review the voltages in all states to make sure they make sense. The Prefault state voltages at each end should have nearly the same magnitudes and angles. The only voltage difference between transmission line ends will be the voltage drop across the transmission line caused by a relatively small current flow.

Review the Prefault state current magnitudes and angles and make sure:

- That the magnitudes are relatively close. The difference between one end and the other will be line losses, which should not be significant.
- The angles are approximately 180° apart. Prefault currents should represent current flowing through a transmission line which means the current entering one side should leave the other side. They may not be exactly 180° apart due to line losses but they should obviously be in opposite directions.

iv) Are Fault State Voltages and Currents Correct?

The faulted voltages (A-N, A-B, 3-P, etc.) in any Fault state should meet the following criteria:

- Have lower magnitudes than the Prefault voltages on the same phases at all ends.
- Include a larger voltage drop on the closest end to the fault unless there is a huge difference between sources at each end.
- The faulted phase(s) of a Phase-to-Neutral or 3-Phase fault will have roughly the same phase angle(s).
- The faulted phases of a Phase-to-Phase fault will have decreased but nearly identical magnitudes and the phasor angles will change to come closer together. If a fault is located directly next to one end, that end's voltages will have nearly identical magnitudes and angles.

The non-faulted voltages in the Fault states should be close to the Prefault magnitudes and angles.

The faulted currents (A-N, A-B, 3-P, etc.) in any Fault state should:

- Have higher magnitudes than the Prefault currents on the same phases at all ends.
- Have a larger current rise on the faulted end, unless there is a huge difference between sources at each end.
- Have similar phase angles between the two sides if the fault is located between the two relays because the source on each side will contribute to the fault and current should flow into the transmission line from both ends.

- Have opposite phase angles between the two sides if the fault is not on the transmission line because one side will feed the fault located beyond the other side. The current will flow into the transmission line through the polarity of one relay and leave the transmission line through the non-polarity of the other relays like the Prefault state currents.

v) Is Postfault Enabled and are the Values Correct?

The Postfault state voltages, if enabled, should simulate the voltages connected to the relay when the transmission line is isolated from the grid. This usually means that the Prefault and Postfault state voltages will be identical, but they could be different if one side is islanded from the other when the transmission line opens.

vi) Are the Data Test Plan Vectors Correct?

Reviewing data plans is much easier than reviewing waveform files because the magnitudes and angles can be:

- Visualized by an experienced relay tester, or
- Displayed by most test-sets as some kind of phasor display.

RLY-1	TEST-SET SECONDARY VALUES					
	PREFAULT		FAULT 1		POSTFAULT	
VA	66.4	0.0	55.5	-1.0	66.4	0.0
VB	66.4	-120.0	55.5	-121.0	66.4	-120.0
VC	66.4	120.0	55.5	119.0	66.4	120.0
IA	0.50	-10.0	11.13	-81.0	0.0	0.0
IB	0.50	-130.0	11.13	159.0	0.0	0.0
IC	0.50	110.0	11.13	39.0	0.0	0.0
Time (cycles)	60.00		10.00		60.0	

RLY-2	TEST-SET SECONDARY VALUES					
	PREFAULT		FAULT 1		POSTFAULT	
VA	66.4	0.0	59.6	-1.0	66.4	0.0
VB	66.4	-120.0	59.6	-121.0	66.4	-120.0
VC	66.4	120.0	59.6	119.0	66.4	120.0
IA	0.33	170.0	7.97	-81.0	0.0	0.0
IB	0.33	50.0	7.97	159.0	0.0	0.0
IC	0.33	-70.0	7.97	39.0	0.0	0.0
Time (cycles)	60.00		10.00		60.0	

Figure 2-9: Example Test Plan Comparison #1

You should be able to determine the following information from the data plan displayed in Figure 2-9.

Prefault

- The voltage settings look correct because the voltages, magnitudes, and angles are the same.
- The current settings appear to be correct because the magnitudes are roughly the same and the angles for each phase appear to be 180° apart. The current appears to be flowing in RLY-1 and out RLY-2.
- The times are identical.

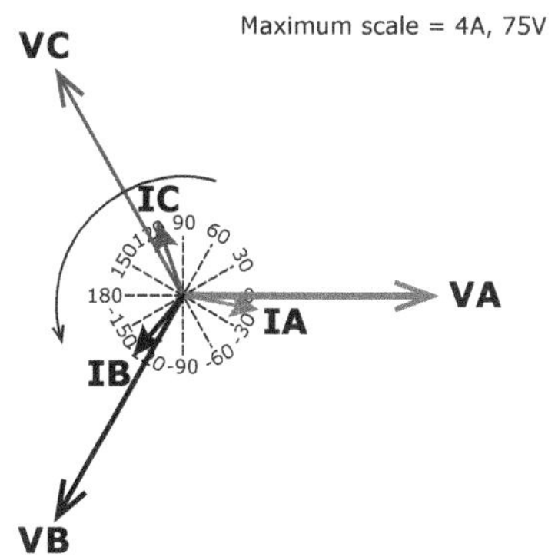

Figure 2-10: RLY-1 Prefault Vectors

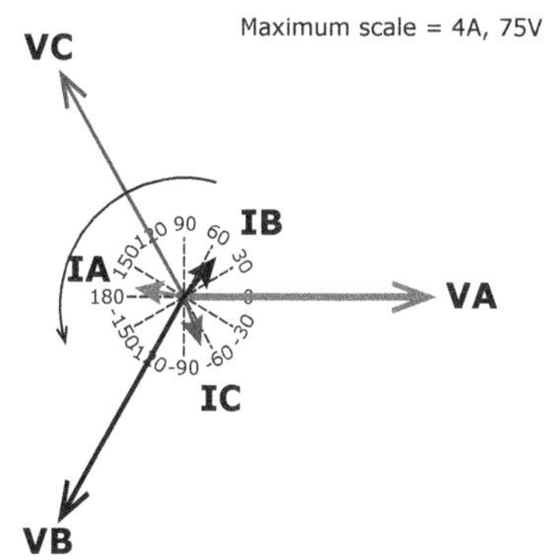

Figure 2-11: RLY-2 Prefault Vectors

Fault 1

- The voltage and current settings at both ends indicate an A-B-C fault.
- The RLY-1 faulted voltage magnitudes have decreased more than RLY-2, so either the fault is closer to RLY-1, or the RLY-1 source is weaker than the source connected to RLY-2.
- All three voltages connected to each relay have equal magnitudes, have remained approximately 120° apart, and are roughly the same between relays; so this is a 3-Phase fault.
- All three RLY-1 currents have equal magnitudes and are roughly 120° apart, so this is a 3-Phase fault.
- The RLY-1 currents are greater than the RLY-2 currents, so the fault is either closer to RLY-1 or the RLY-1 source is stronger.
- All three RLY-2 currents have equal magnitudes and are roughly 120° apart, so this is a 3-Phase fault.
- The RLY-1 and RLY-2 faulted phases have roughly the same angles, so the fault is located on the line.
- The times are identical.

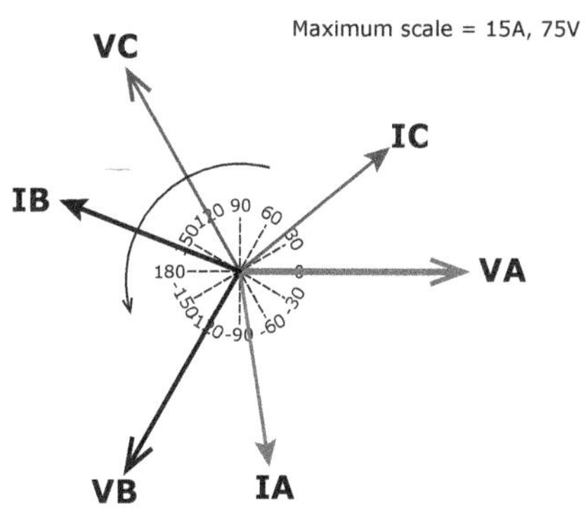

Figure 2-12: RLY-1 Fault 1 Vectors

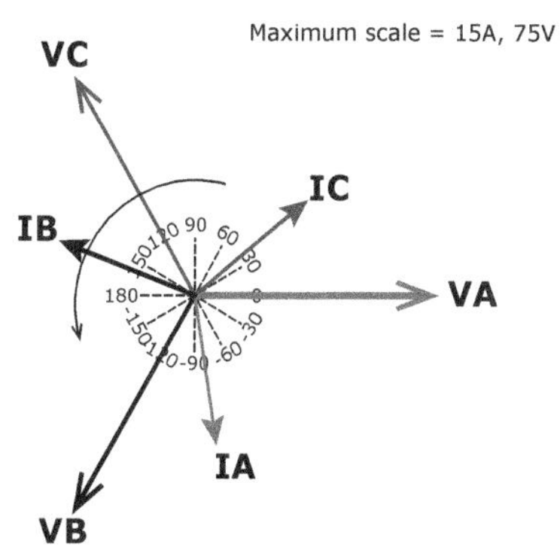

Figure 2-13: RLY-2 Fault 1 Vectors

Postfault

- The voltage settings look correct because the voltages, magnitudes, and angles are the same.
- There is no current, so the breakers must be open.
- The times are identical.

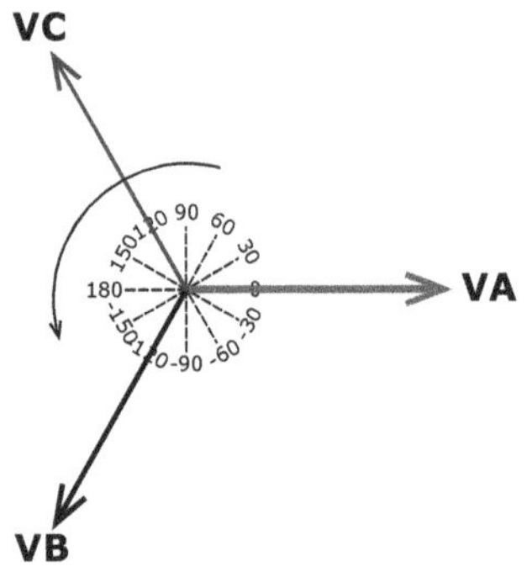

Figure 2-14: RLY-1 Postfault Vectors

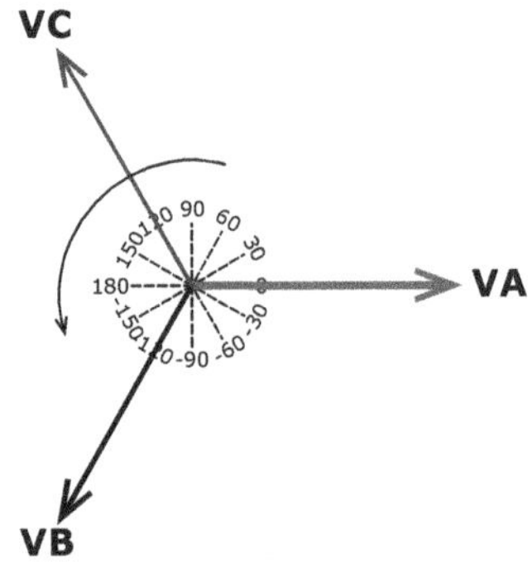

Figure 2-15: RLY-2 Postfault Vectors

Here's another data plan.

RLY-1	TEST-SET SECONDARY VALUES					
	PREFAULT		FAULT 1		POSTFAULT	
VA	66.4	0.0	28.9	-4.0	66.4	0.0
VB	66.4	-120.0	69.5	-123.0	66.4	-120.0
VC	66.4	120.0	69.0	123.0	66.4	120.0
IA	0.50	-10.0	37.99	-83.0	0.00	0.0
IB	0.50	-130.0	0.59	-89.0	0.00	0.0
IC	0.50	110.0	0.59	-79.0	0.00	0.0
Time (cycles)	60.00		10.00		60.0	

RLY-2	TEST-SET SECONDARY VALUES					
	PREFAULT		FAULT 1		POSTFAULT	
VA	66.4	0.0	60.1	-1.0	66.4	0.0
VB	66.4	-120.0	66.1	-119.0	66.4	-120.0
VC	66.4	120.0	66.2	119.0	66.4	120.0
IA	0.33	170.0	2.76	-80.0	0.00	0.0
IB	0.33	50.0	0.39	91.0	0.00	0.0
IC	0.33	-70.0	0.39	101.0	0.00	0.0
Time (cycles)	60.00		10.00		60.0	

Figure 2-16: Example Test Plan Comparison #2

You should be able to determine the following information from the data plan displayed in Figure 2-16.

Prefault

- The voltage settings look correct because the voltages, magnitudes, and angles are the same.
- The current settings appear to be correct because the magnitudes are roughly the same and the angles for each phase appear to be 180° apart. The current appears to be flowing in RLY-1 and out RLY-2.
- The times are identical.

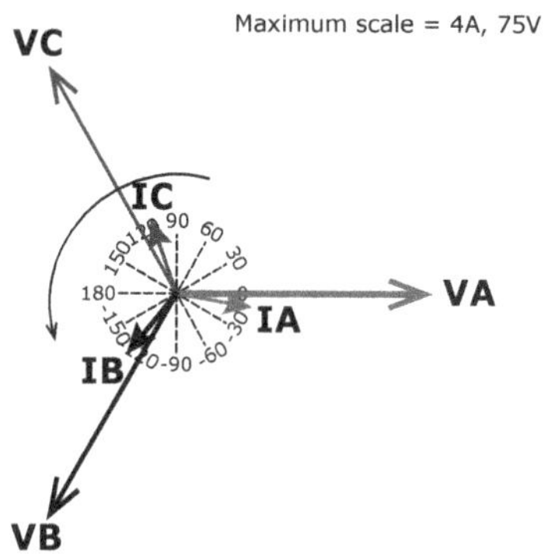

Figure 2-17: RLY-1 Prefault Vectors

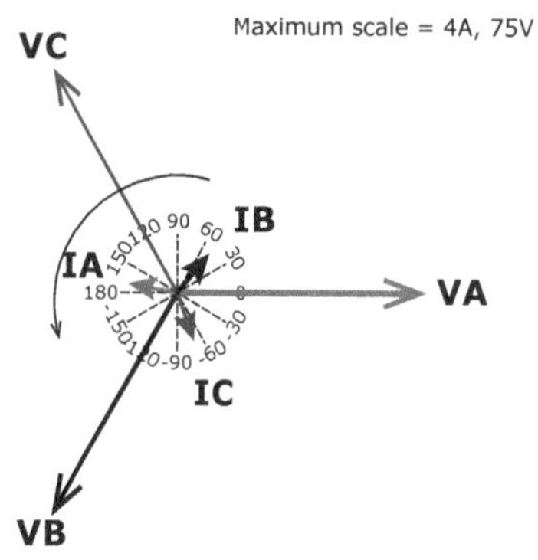

Figure 2-18: RLY-2 Prefault Vectors

Fault 1

- The voltage and current settings on both ends indicate an A-N fault.
- The RLY-1 faulted voltage magnitudes have decreased more than RLY-2, so either the fault is closer to RLY-1 or the RLY-1 source is weaker than the source connected to RLY-2.
- The RLY-1 fault current is greater than the RLY-2 current, so the fault is either closer to RLY-1 or the RLY-1 source is stronger.
- The RLY-1 and RLY-2 faulted phases have roughly the same angles, so the fault is located on the line.
- The times are identical.

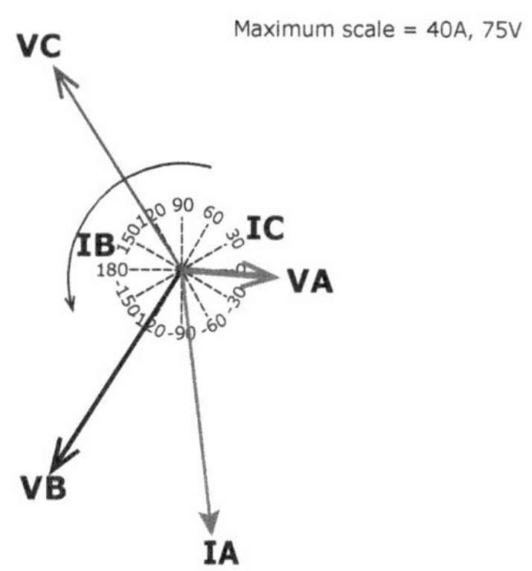

Figure 2-19: RLY-1 Fault 1 Vectors

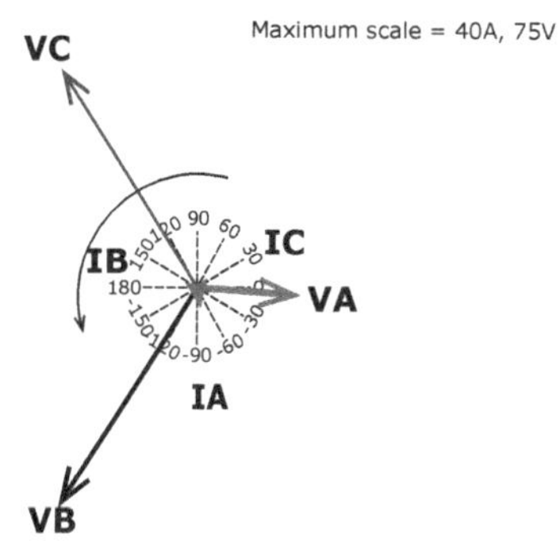

Figure 2-20: RLY-2 Fault 1 Vectors

Postfault

- The voltage settings look correct because the voltages, magnitudes and angles are the same.
- There is no current, so the breakers must be open.
- The times are identical.

The next example is a little more complicated because it is a phase-reversal test where a fault behind RLY-1 is simulated in Fault 1 and then the current suddenly changes direction and the fault is located behind RLY-2.

RLY-1	Test Case #1							
	PREFAULT		FAULT 1		FAULT 2		Post Fault	
VA	66.4	0.0	67.3	0.0	67.3	0.0	66.4	0.0
VB	66.4	-120.0	33.7	180.0	52.5	-130.0	66.4	-120.0
VC	66.4	120.0	33.7	180.0	52.3	129.0	66.4	120.0
IA	0.50	-10.0	0.06	-85.0	0.06	-85.0	0.0	-10.0
IB	0.50	-130.0	4.65	9.0	3.66	-171.0	0.0	-130.0
IC	0.50	110.0	4.65	-172.0	3.66	10.0	0.0	110.0
Time (cy)	60.00		10.00		10.00		60.00	

RLY-2	Test Case #2							
	PREFAULT		FAULT 1		FAULT 2		Post Fault	
VA	66.4	0.0	66.7	0.0	66.7	0.0	66.4	0.0
VB	66.4	-120.0	57.5	-126.0	34.1	-173.0	66.4	-120.0
VC	66.4	120.0	56.7	125.0	33.1	173.0	66.4	120.0
IA	0.33	170.0	0.04	95.0	0.04	95.0	0.0	170.0
IB	0.33	50.0	3.10	-171.0	2.44	9.0	0.0	50.0
IC	0.33	-70.0	3.10	8.0	2.44	-170.0	0.0	-70.0
Time (cy)	60.00		10.00		10.00		60.00	

Figure 2-21: Example Test Plan Comparison

You should be able to determine the following information from the data plan displayed in Figure 2-21.

Prefault

- The voltage settings look correct because the voltages, magnitudes, and angles are the same.
- The current settings appear to be correct because the magnitudes are roughly the same and the angles for each phase appear to be 180° apart. The current appears to be flowing in RLY-1 and out RLY-2.
- The times are identical.

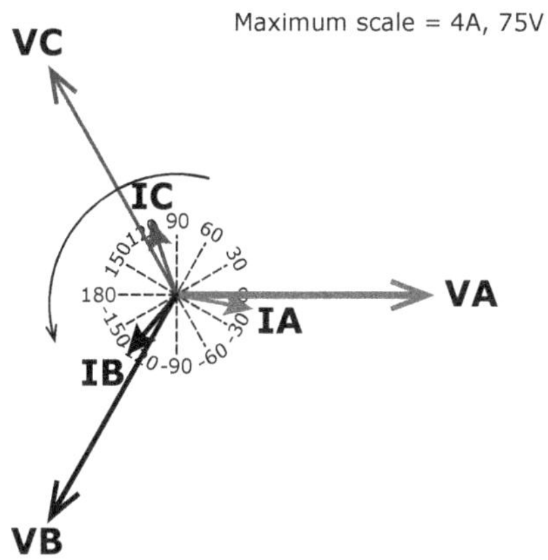

Figure 2-22: RLY-1 Prefault Vectors

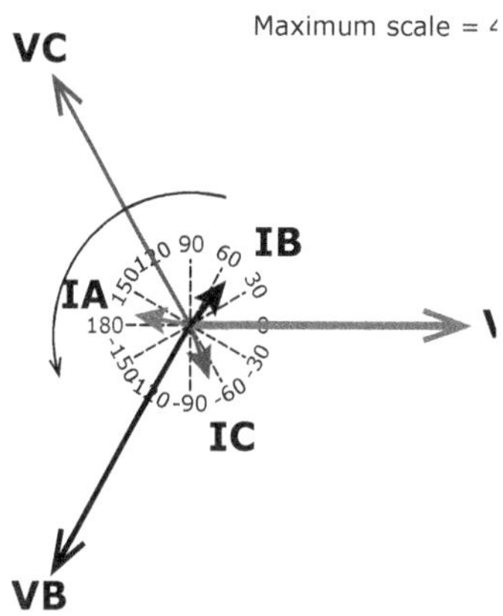

Figure 2-23: RLY-2 Prefault Vectors

Fault 1

- The voltage and current settings on both ends indicate a B-C fault.
- The non-faulted voltage (VA) doesn't change very much from Prefault to Fault.
- The RLY-1 faulted voltage magnitudes and angles are equal, so the fault is very close to this relay's PTs.
- The RLY-2 faulted voltage magnitudes have decreased and the angles have come together as they should during a P-P fault.
- The RLY-1 B and C-Phase currents are roughly equal and opposite as they should be during a P-P fault. The IB current is in the opposite quadrant when compared to the Prefault current, so the fault is behind this relay.
- The RLY-2 B and C-Phase currents are roughly equal and opposite as they should be during a P-P fault. The IB current is opposite the RLY-1 IB current, so the fault is not on the transmission line.
- The times are identical.

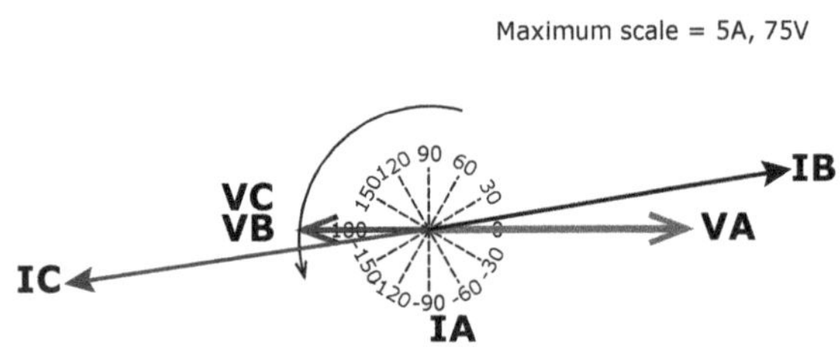

Figure 2-24: RLY-1 Fault 1 Vectors

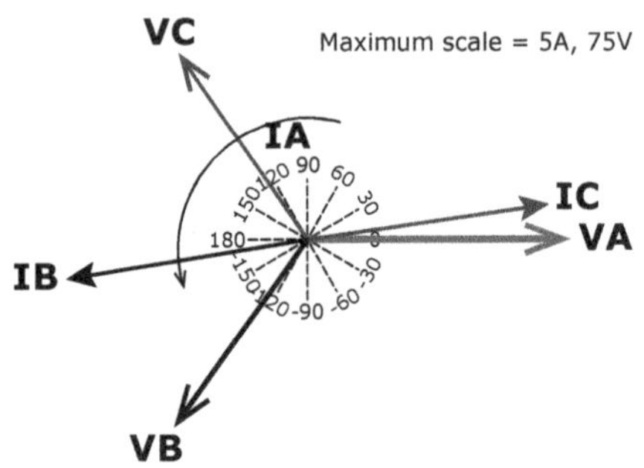

Figure 2-25: RLY-2 Fault 1 Vectors

The example test plan is copied here to help you compare values.

RLY-1	Test Case #1							
	PREFAULT		FAULT 1		FAULT 2		Post Fault	
VA	66.4	0.0	67.3	0.0	67.3	0.0	66.4	0.0
VB	66.4	-120.0	33.7	180.0	52.5	-130.0	66.4	-120.0
VC	66.4	120.0	33.7	180.0	52.3	129.0	66.4	120.0
IA	0.50	-10.0	0.06	-85.0	0.06	-85.0	0.0	-10.0
IB	0.50	-130.0	4.65	9.0	3.66	-171.0	0.0	-130.0
IC	0.50	110.0	4.65	-172.0	3.66	10.0	0.0	110.0
Time (cy)	60.00		10.00		10.00		60.00	

RLY-2	Test Case #2							
	PREFAULT		FAULT 1		FAULT 2		Post Fault	
VA	66.4	0.0	66.7	0.0	66.7	0.0	66.4	0.0
VB	66.4	-120.0	57.5	-126.0	34.1	-173.0	66.4	-120.0
VC	66.4	120.0	56.7	125.0	33.1	173.0	66.4	120.0
IA	0.33	170.0	0.04	95.0	0.04	95.0	0.0	170.0
IB	0.33	50.0	3.10	-171.0	2.44	9.0	0.0	50.0
IC	0.33	-70.0	3.10	8.0	2.44	-170.0	0.0	-70.0
Time (cy)	60.00		10.00		10.00		60.00	

Fault 2

- The voltage and current settings on both ends indicate a B-C fault.
- The non-faulted voltage (VA) doesn't change very much from Prefault to Fault.
- The RLY-2 faulted voltage magnitudes and angles are close together, so the fault is closer to this relay's PTs.
- The RLY-1 faulted voltage magnitudes have decreased and the angles have come together as they should during a P-P fault.
- The RLY-2 B and C-Phase currents are roughly equal and opposite as they should be during a P-P fault. The IB current is in the same quadrant when compared to the Prefault current, so the fault is behind this relay.
- The RLY-1 B and C-Phase currents are roughly equal and opposite as they should be during a P-P fault. The IB current is opposite the RLY-2 IB current, so the fault is not on the transmission line.
- The times are identical.

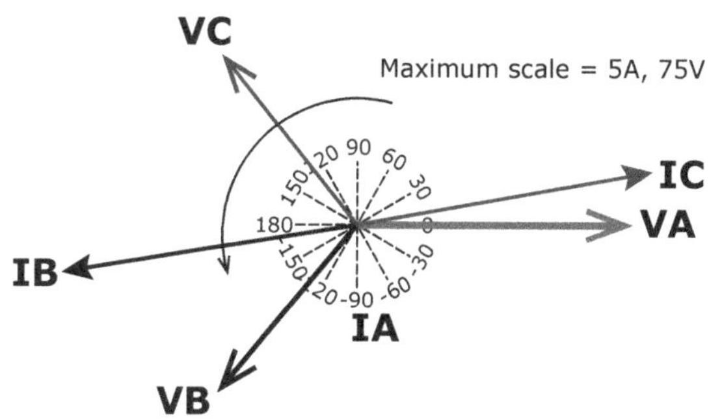

Figure 2-26: RLY-1 Fault 2 Vectors

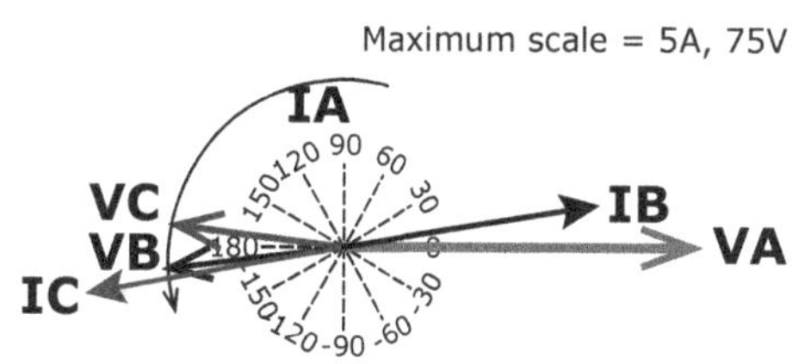

Figure 2-27: RLY-2 Fault 2 Vectors

Postfault

- The voltage settings look correct because the voltages, magnitudes, and angles are the same.
- There is no current, so the breakers must be open.
- The times are identical.

vii) Are the Waveform Test Plan Vectors Correct?

Waveforms are much more difficult to check and every software program has different controls and/or features, so we will only discuss the basics in this section using the examples in Figures 2-28, 2-29, and 2-30.

RLY-1	TEST-SET SECONDARY VALUES					
	PREFAULT		FAULT 1		POSTFAULT	
VA	66.4	0.0	28.9	-4.0	66.4	0.0
VB	66.4	-120.0	69.5	-123.0	66.4	-120.0
VC	66.4	120.0	69.0	123.0	66.4	120.0
IA	0.50	-10.0	37.99	-83.0	0.00	0.0
IB	0.50	-130.0	0.59	-89.0	0.00	0.0
IC	0.50	110.0	0.59	-79.0	0.00	0.0
Time (cycles)	60.00		10.00		60.0	

RLY-2	TEST-SET SECONDARY VALUES					
	PREFAULT		FAULT 1		POSTFAULT	
VA	66.4	0.0	60.1	-1.0	66.4	0.0
VB	66.4	-120.0	66.1	-119.0	66.4	-120.0
VC	66.4	120.0	66.2	119.0	66.4	120.0
IA	0.33	170.0	2.76	-80.0	0.00	0.0
IB	0.33	50.0	0.39	91.0	0.00	0.0
IC	0.33	-70.0	0.39	101.0	0.00	0.0
Time (cycles)	60.00		10.00		60.0	

Figure 2-28: Example Test Plan 2 with Raw Data

You would not normally get the information from Figure 2-28, so we will only look at the two waveforms (Figures 2-29 and 2-30) on a split screen to try and find any obvious mistakes.

Prefault

- The voltages look correct because the voltage magnitudes appear to be equal.
- There appears to be Prefault current and we would have to zoom in to see if the magnitudes were similar.
- The Prefault times appear to be equal in both waveforms at approximately 0.5 seconds.

Fault 1

- The VA voltage appears to decrease in both waveforms and the IA currents are higher, so this appears to be an A-N fault.
- The RLY-1 faulted voltage magnitudes have decreased more than RLY-2, so either the fault is closer to RLY-1 or the RLY-1 source is weaker than the source connected to RLY-2.
- The RLY-1 fault current is greater than the RLY-2 currents, so the fault is either closer to RLY-1 or the RLY-1 source is stronger.
- The Fault times appear to be identical.

Postfault

- The voltages look correct because the voltage magnitudes are the same.
- There is no current, so the breakers must be open.
- The times are identical.

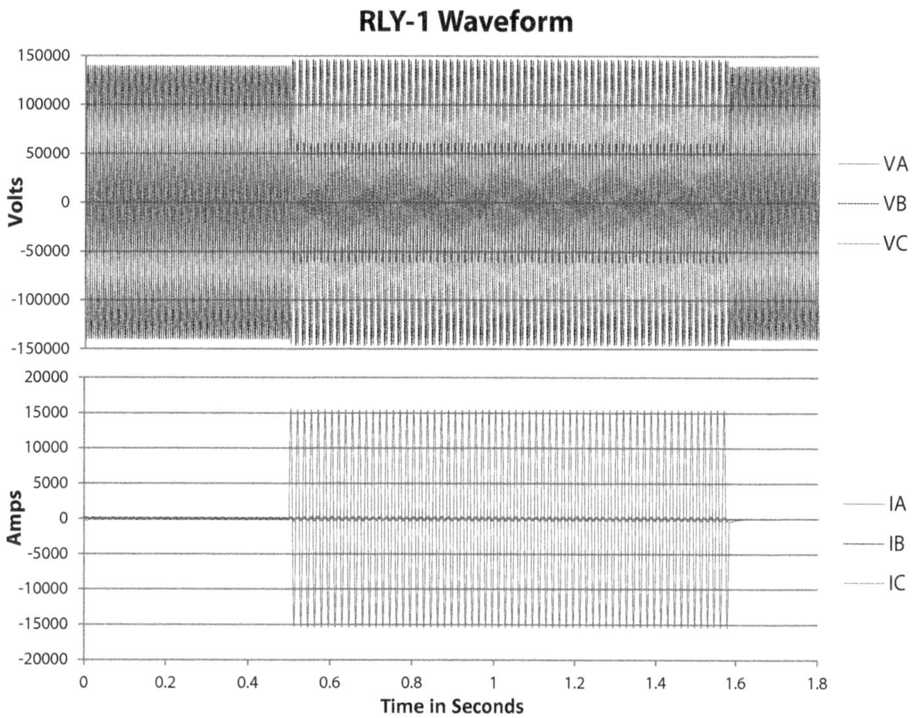

Figure 2-29: Example RLY-1 Waveform

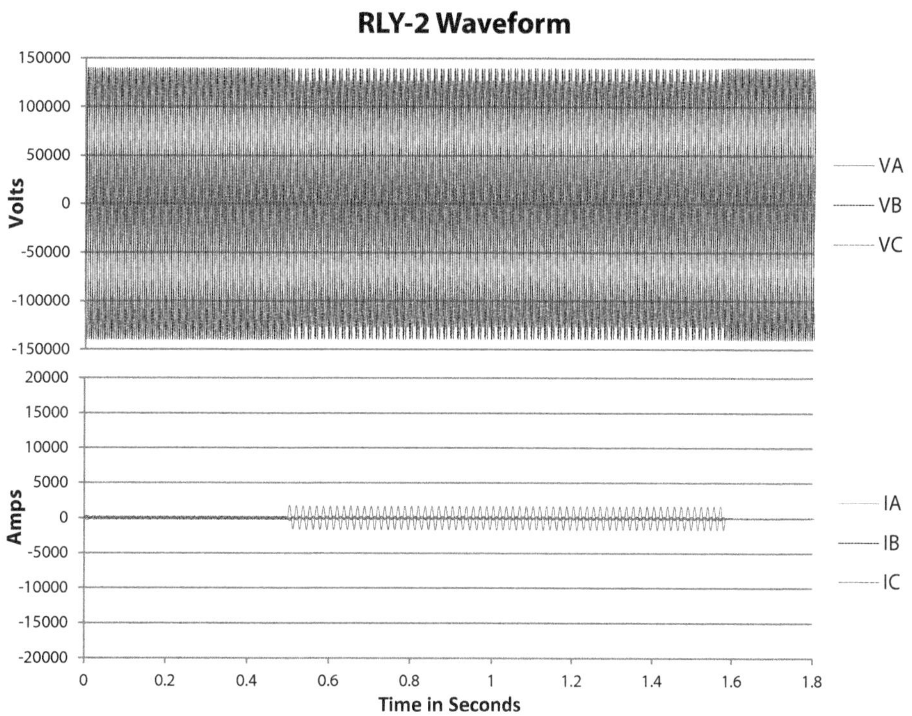

Figure 2-30: Example RLY-2 Waveform

It is difficult to determine if the angles are correct without zooming in and selecting a reference at the exact same time in both waveforms as shown in the following figures. Figure 2-31 zooms into the Prefault voltage waveforms with an arbitrary line drawn between the two relays at 0.01 seconds for comparison. Each phase should have nearly identical values as per the following analysis:

- The A-Phase voltages have just finished their negative peak before 0.01 seconds, so the A-Phase voltages are in-phase.
- The B-Phase voltages appear to be at the zero crossing between positive and negative peaks, so the B-Phase voltages appear to be in-phase.
- The C-Phase voltage's positive peaks are just about to occur in both waveform diagrams at 0.01 seconds, so the C-Phase vectors appear to be in-phase.

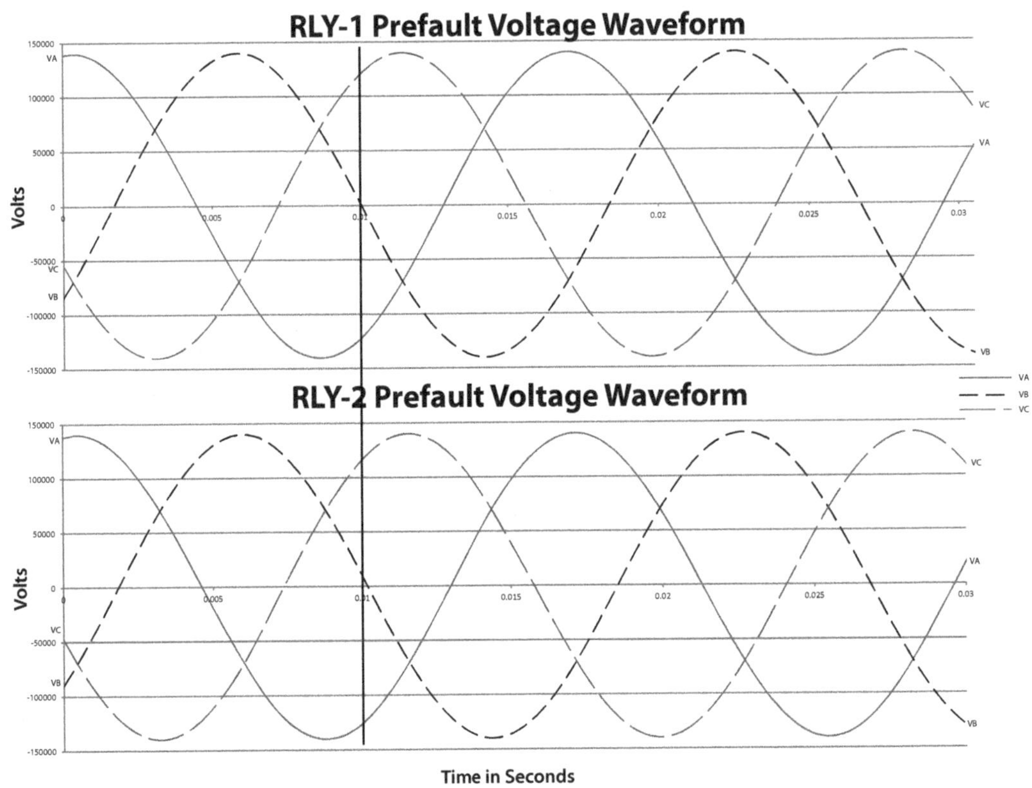

Figure 2-31: Example Prefault Voltage Waveform Comparison

Figure 2-32 zooms into the Prefault current waveforms with an arbitrary line drawn between the two relays at 0.01 seconds for comparison. Each phase should have opposite values as per the following analysis:

- The RLY-1 A-Phase current just finished its negative peak and the RLY-2 A-Phase current just finished its positive peak before 0.01 seconds. The A-Phase currents are opposite, so the A-Phase current enters RLY-1 and leaves RLY-2.
- The RLY-1 B-Phase current appears to be near the zero crossing between positive and negative peaks and the RLY-2 B-Phase current appears to be near the zero crossing between negative and positive peaks. The currents are opposite so the B-Phase current enters RLY-1 and leaves RLY-2.
- The RLY-1 C-Phase current appears to be near the positive peak, and the RLY-2 C-Phase current appears to be near the negative peak. The currents are opposite, so the C-Phase current enters RLY-1 and leaves RLY-2.

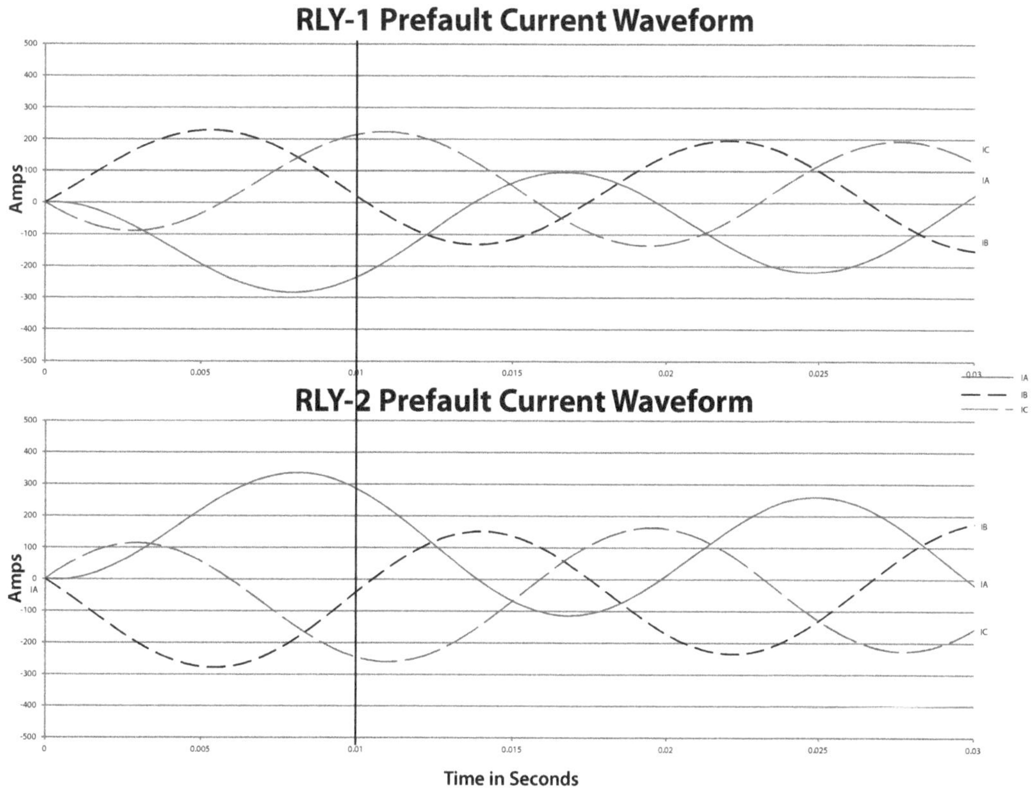

Figure 2-32: Example Prefault Current Waveform Comparison

Figure 2-33 zooms into the Fault voltage waveforms with an arbitrary line drawn between the two relays at 0.51 seconds for analysis:

- The A-Phase voltages just finished their negative peak before 0.51 seconds, so the A-Phase voltages are in-phase. The RLY-1 voltage is significantly less than the RLY-2 voltage, so the fault is likely closer to RLY-1.
- The B-Phase voltages appear to be at the zero crossing between positive and negative peaks, so the B-Phase voltages appear to be in-phase.
- The C-Phase voltage's positive peaks are just about to occur in both waveform diagrams at 0.51 seconds, so the C-Phase vectors appear to be in-phase.

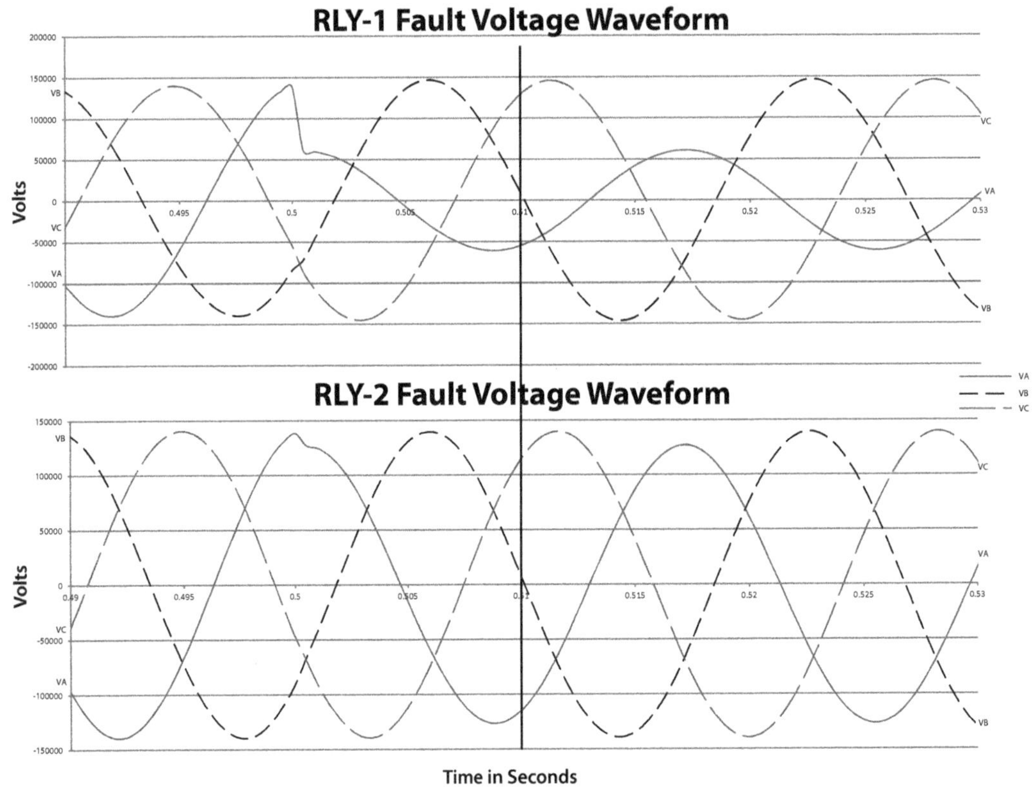

Figure 2-33: Example Fault Voltage Waveform Comparison

Figure 2-34 zooms into the Fault current waveforms with an arbitrary line drawn between the two relays at 0.51 seconds for analysis:

- Both A-Phase currents appear to be heading to their negative peaks, so the currents are in-phase and the fault is between the two relays.
- The waveform scaling makes it difficult to see, but the B and C-Phase waveforms have similar magnitudes and the peaks are opposite, so it appears that the B and C phases are correct as well.

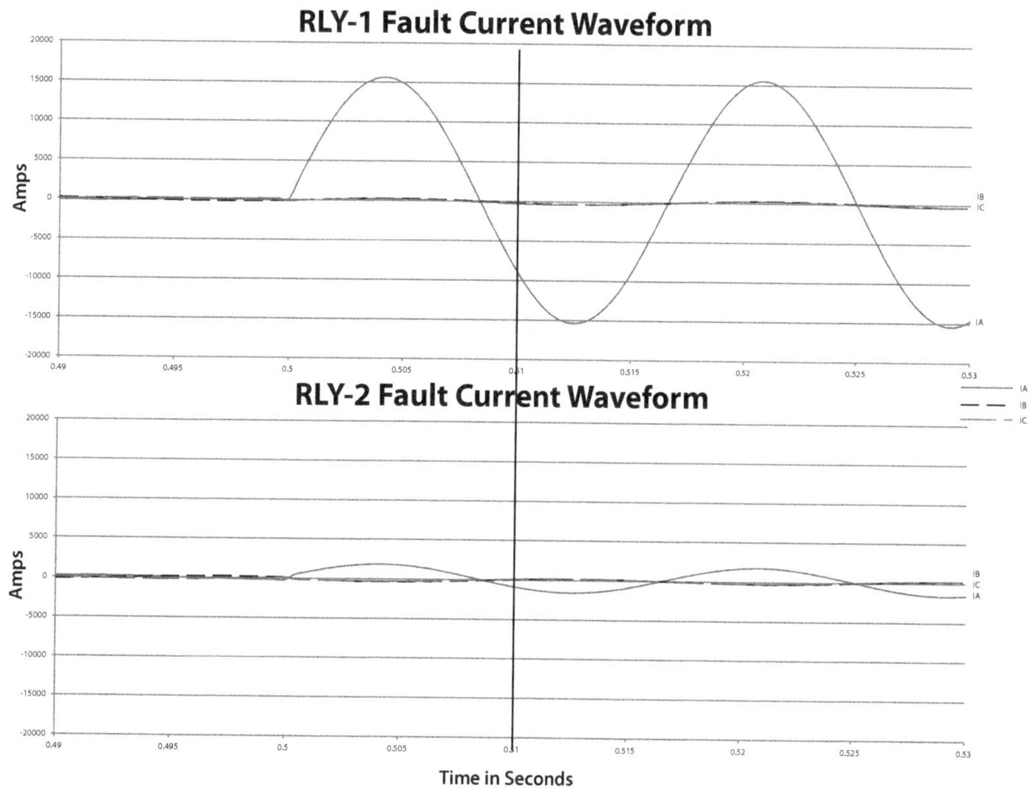

Figure 2-34: Example Fault Current Waveform Comparison

2. Set up the GPS Antenna

Most test-sets require at least 4 GPS satellite signals to guarantee accuracy. The antenna must be mounted in a fairly open area with a clear view of the open sky. Mounting the antenna on top of a truck in the parking lot usually works well. (As long as you don't drive away for lunch with it still attached) Change the test-set settings to use the GPS clock for its reference signal and wait for the GPS status to indicate that the test-set has been synchronized to GPS. This can take up to 30 minutes the first time it is applied at a location depending on the test-set, so it is always a good idea to perform this step first. Subsequent synchronizations shouldn't require such a long delay. Become familiar with GPS error messages by disconnecting the test-set from the antenna after GPS synchronization in order to recognize a loss-of-synchronization problem if it occurs during the test procedure.

If open sky is not available, some test-sets will allow synchronization to the substation's IRIG signal. A substation's IRIG signal is another timing standard that uses a GPS clock connected to a large antenna that converts the GPS time to an IRIG signal. The IRIG signal is connected to all of the protective relays, fault recorders, and other devices inside the substation to ensure all devices will record the same time if an event occurs to help with post-fault analysis.

3. Isolate the Equipment Under Test

An ideal end-to-end test requires the transmission line to be completely isolated by disconnect switches outside the zone of protection. This allows the circuit breakers to operate in order to prove the entire protective scheme as shown in the following figure.

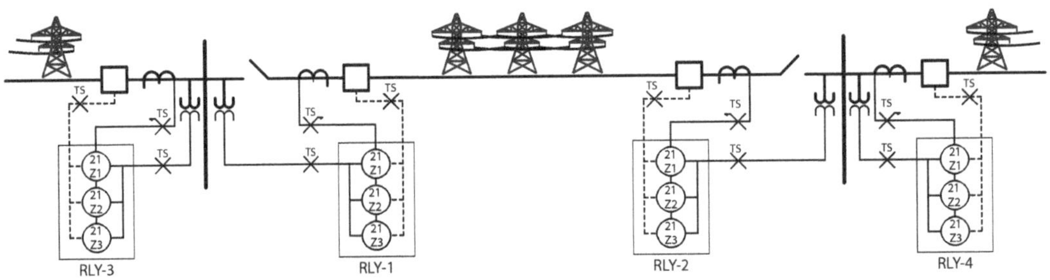

Figure 2-35: High Voltage Isolation

It is possible to perform end-to-end tests without isolating the line, but special care must be taken to ensure that the circuit breakers remain closed throughout the test and that backup protection systems are available. Online end-to-end testing is performed by isolating the primary protection via test switches while the backup protection remains online to provide protection for the transmission line while the tests are performed. Simulating breaker status contacts is required and often difficult, as test switches may not be available for breaker status inputs to the relay, depending on the location and local utility standards. After the primary protection is tested, it is carefully placed back into service and the process is repeated for testing the backup protection.

4. Relay Input and Output Connections

Carefully review the drawings to make sure all output contacts are accounted for and open any test switches, panel circuit breakers, fuses, etc. necessary to prevent unintended equipment operation. The circuit breaker position, relay operation, or metering values applied during testing could have unforeseen consequences in an external plant-wide logic controller causing embarrassing and expensive outages if appropriate measures are not taken.

There are several ways to connect relay output contacts to the test-set depending on the test-set and the field connections. The simplest connection applies the test-set input contacts directly across the relay's output contacts. With test switches, this is a simple task as shown in Figure 2-36. TS-52-5-DC1 switches 1-2 and 3-4 are opened and the test-set input is connected at the test switches or on the relay itself. Test switches are nice but not always available, and a test-set input can be connected across the contact without test switches as shown on the right side of Figure 2-36. Check with the test-set manufacturer before attempting this connection. Some relay manufacturer inputs are polarity sensitive and may need to be reversed if the test-set senses the contact is closed when it is actually open. If the circuit breakers will operate during the test, the test switches should be closed to allow the trip signal to operate the circuit breaker's trip coil. Open the test switches if the breaker is not intended to operate during the test.

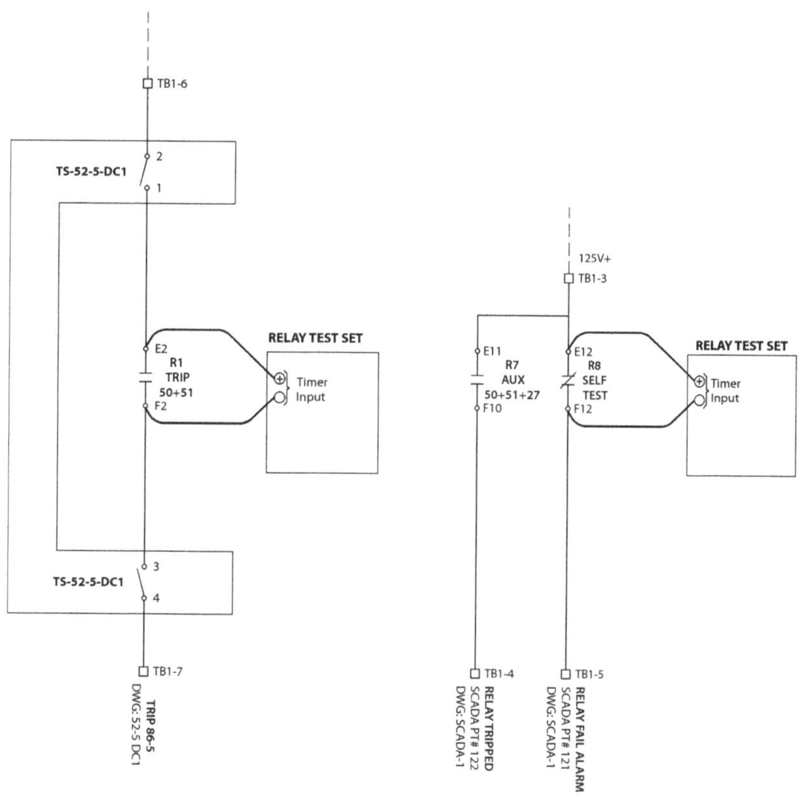

Figure 2-36: Simple Test-Set Input Connections

If test switches are not provided or are closed, another part of the circuit could be shorted in parallel with the output contact under test and cause a false operation. The relay "R2 Close PB" contact in the following figures is connected in parallel with the "DCS close" contact. If the DCS contact closes when the test switches are closed, the relay input will sense contact closure. This problem is easily solved by opening either of the test switches, if available. One wire must be removed when no test switches are provided. Figure 2-37 displays the different options when contacts are connected in parallel.

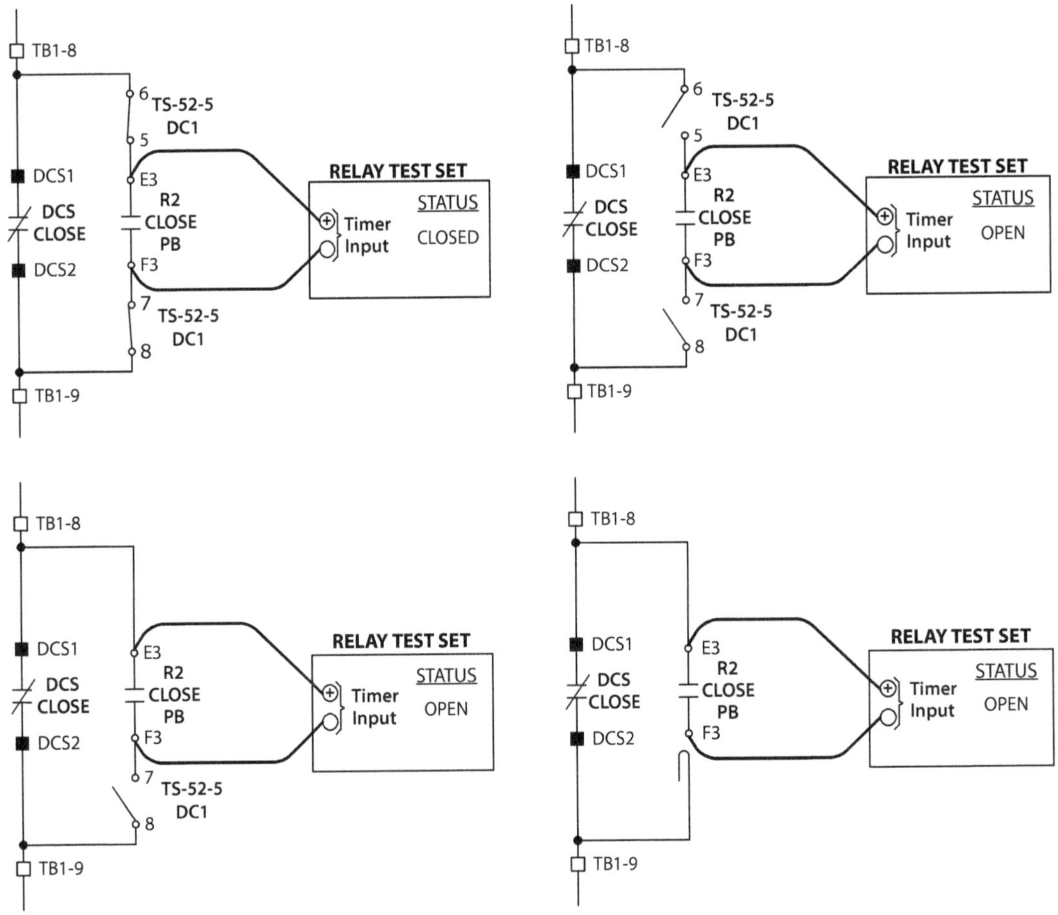

Figure 2-37: Test-Set Input Connections with Contact in Parallel

Most test-set manufacturers also allow voltage-monitoring inputs to reduce wiring changes when testing. Instead of monitoring whether a contact is closed or open, the voltage-monitoring option determines that the contact is closed when the measured voltage is above the test-set's defined setpoint. The test-set assumes the contact is open if the measured voltage is below the setpoint. Another connection is required when using voltage-detecting test-set inputs. Any of the test-set connections in Figure 2-38 can be used when voltage is required for contact sensing.

Starting from the left, the R2 timer is connected between TB1-9 and TB3-6 (negative circuit). When R2 and "DCS close" are open, the voltage between the two terminals should be negligible and the relay will assume an open contact. When R2 or "DCS close" closes, the relay will detect $125V_{DC}$ across the contacts and the test-set will detect a closed contact. Be wary of this connection because the circuit breaker will close if the circuit is complete!

The R3 timer is connected between terminal 11 of the open test switch (TS-52-5-DC1) and circuit negative. This is a safe connection as the test-set will detect the correct contact position and the circuit breaker will not trip when the contact is operated.

The simplest connection is the R1 timer with the test-set input connected between terminal 3 of test switch (TS-52-5-DC1) and ground. (The test switch cover screw is the ground that works in most applications.) Obviously this connection will only work when the DC system is grounded at the midpoint, as most DC systems are. This connection is also safe as the connected 86-5 lockout will not operate when the contact closes.

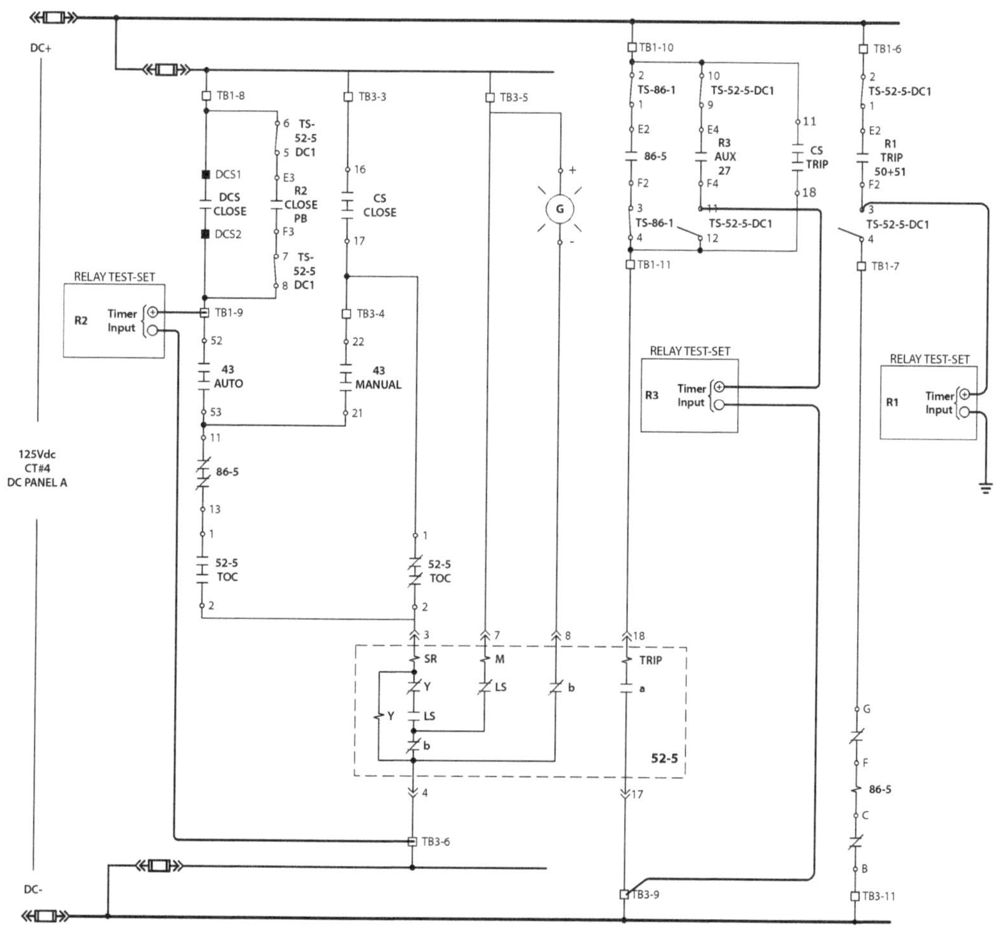

Figure 2-38: Test-Set Input Connections in DC Circuit

NEVER apply the following connection in a trip circuit unless there will be no negative results if the circuit is completed and operates. Some test-set sensing contacts have a low impedance and will complete the circuit, causing the end-device to operate.

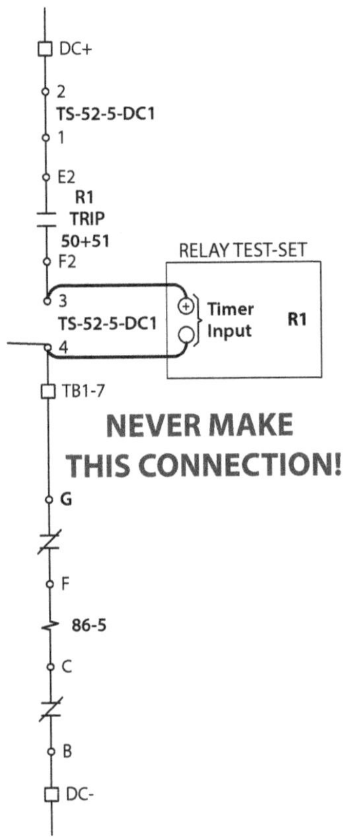

Figure 2-39: Dangerous Test-Set Input Connection in Trip Circuit

Always review the manufacturer's literature when performing digital input testing because relays can be unforgiving when incorrectly connected and cause some embarrassing and expensive smoke to be released. These connections should also be carefully compared to the application to ensure they are connected properly before applying voltage to the circuit. Figures 2-40, 2-41, and 2-42 show some typical examples of input connections from different relay manufacturers.

Figure 2-40 from the Beckwith Electric M-3310 manufacturer's bulletin displays the connections for relay input connections. The field input contact is "dry" and the sensing voltage is supplied by the relay itself. Any external voltage connected in this circuit could damage the relay. The test-set dry output contact or jumper would be connected between terminals 10 and 11 to simulate an IN 1 input.

Figure 2-41 from the SEL-311C manufacturer's bulletin shows that this relay requires "wet" inputs. An external voltage must be connected before the relay will detect input operation. Always check the input voltage to make sure it matches the applied voltage.

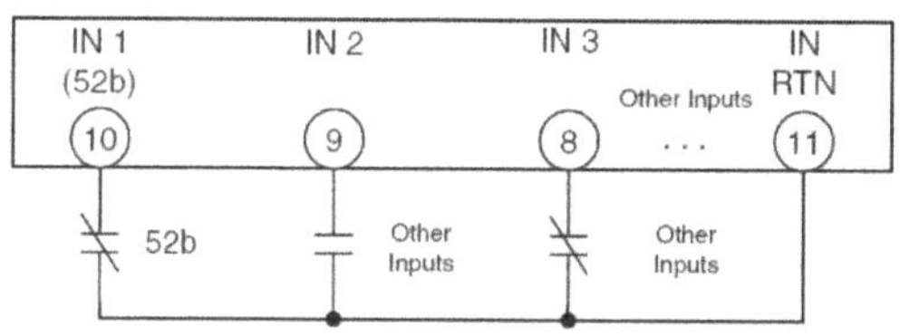

Figure 2-40: M-3310 Relay Input Connections

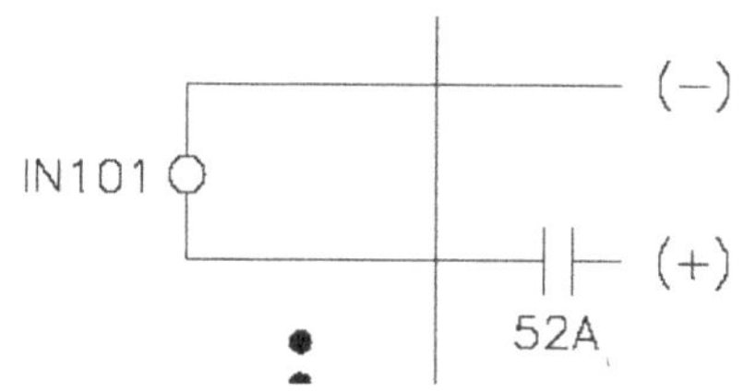

Figure 2-41: SEL 311C Input Connections

GE Multilin SR-750 relays can have "wet" or "dry" contacts connected as shown in Figure 2-42 from the manufacturer's bulletin. Relays that can accept both styles of input contacts are more prone to connection errors. The site and manufacturer's drawings should be compared to ensure no errors have been made.

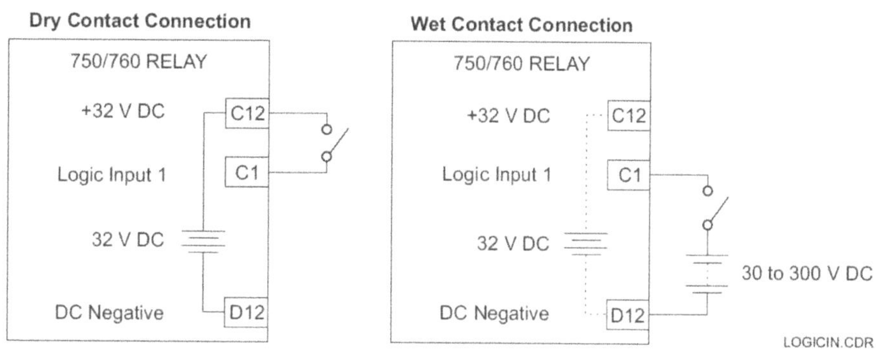

Figure 2-42: GE Multilin SR-750 Input Connections

When testing, the test-set output should be connected across the actual contact used in the circuit to prevent unintentional damage. If the in-service contact is closed, the contact needs to be isolated by opening test switches or temporarily removing wiring in order to test both input states.

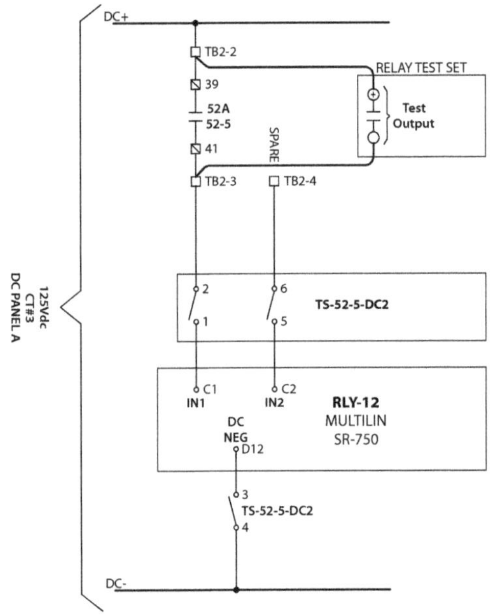

Figure 2-43: Test-Set Output Connections in DC Circuit

5. Connect Test Equipment to Replace CTs/PTs

Connect the test-set to simulate the CT and PT inputs as shown in Figure 2-44. All CT test switches have been opened to short the CT inputs, and an isolating device has been inserted between the CT clips to isolate the top from the bottom.

Always pay attention to the PT connections and triple check that the test-set is connected to the relay side of the test switch. Incorrect connections could back-feed the connected PTs and apply a dangerous voltage to the PTs; or live PTs could damage your test-set.

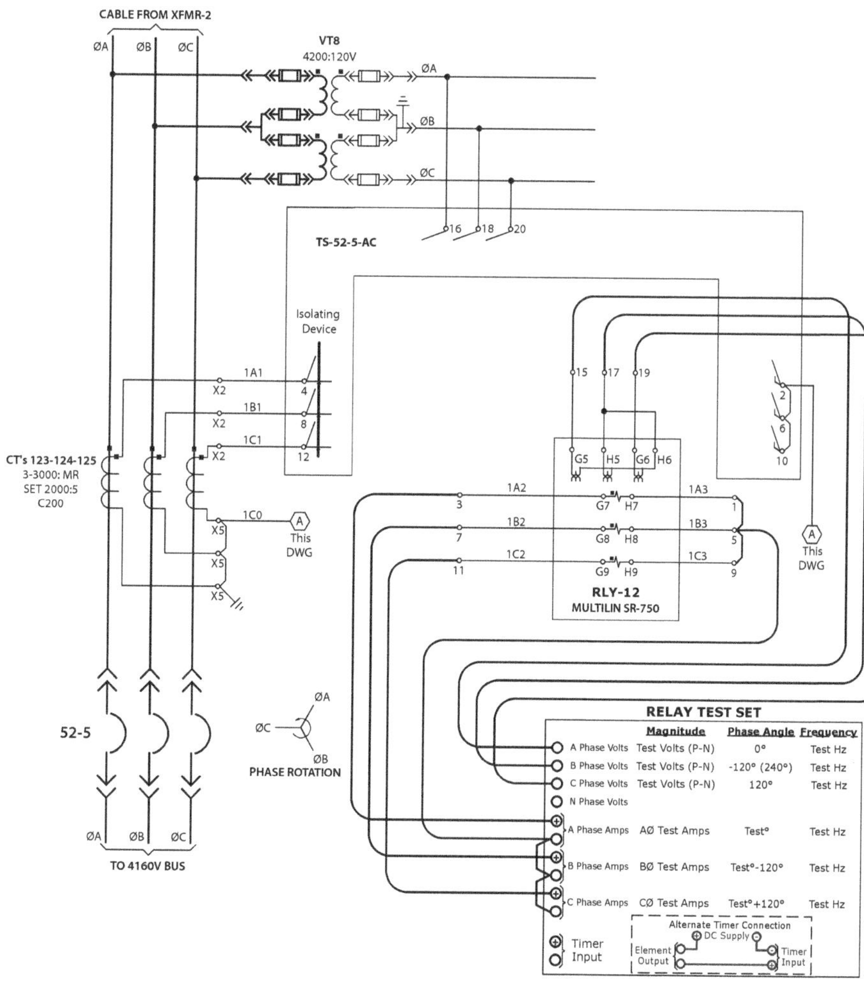

Figure 2-44: Example AC Test-Set Connections

If test switches are not available, the wiring will have to be removed to test the relay. Label each wire and document all connections before removing any wiring. Replace the wiring after testing, and check the terminations against the documentation. Always check the online metering after energization to ensure the wiring has been replaced correctly. It is a good idea to carry a checklist of each wire removed to ensure every wire is returned to service.

6. Apply a Meter Test

A meter test should be the first test performed whenever a digital relay is tested. Meter tests prove that the analog-to-digital converters are working inside the relay, the CT and PT ratios have been set up correctly, and that the test-set to relay connections are correct. It is important to notify all nodes before starting a meter test because your meter test could cause all relays to trip. If the other sides are not isolated, you could cause a system disruption.

Apply single-phase, nominal current and voltage to the relay. Monitor the relay's metering function from the front panel or relay software and ensure that the voltage and current are measured on the correct phase. Apply current and voltage to another phase and ensure that the correct phases are displayed. Repeat for the third phase to make sure all three phases are operating correctly. If the relay monitors zero-sequence voltage and/or current, record the zero-sequence values on the test sheet. When single-phase current/voltage is applied, the zero-sequence value should match the applied value. Zero-sequence components cannot occur in delta connected systems, and there will be no zero-sequence measurements for delta connected PTs.

Apply three-phase, nominal current and voltage to the relay, record the metering results on the test sheet, and compare them to the CT and PT ratios. If the relay also displays phase angles, record these values and ensure that they are in the correct phase relationship. Do not assume that a three-phase, phase-angle measurement can be used in place of the previous single-phase tests. The relay uses its own reference for phase angles which can be misleading. For example, if all three phases were rolled to the next position (AØ to BØ, BØ to CØ, CØ to AØ), the test-set and the relay would both indicate the correct phase angles for each phase (AØ=0°, BØ=-120°, CØ=120°), but AØ current/voltage from the test-set would be injected into BØ of the relay. Also, the test-set and the relay could use different references when displaying phase angles as shown in Figure 2-45, which can add confusion. For example, the phase relationships displayed by a GE Multilin SR-750 would be 0°, 120°, 240° LAG. A SEL relay with the same settings and connections would display 0°, -120°, 120°. If the relay monitors positive-sequence components, record the current and voltage values on the test sheet. The positive-sequence value should match the applied current and voltage, and the negative-sequence and zero-sequence voltages should be almost zero.

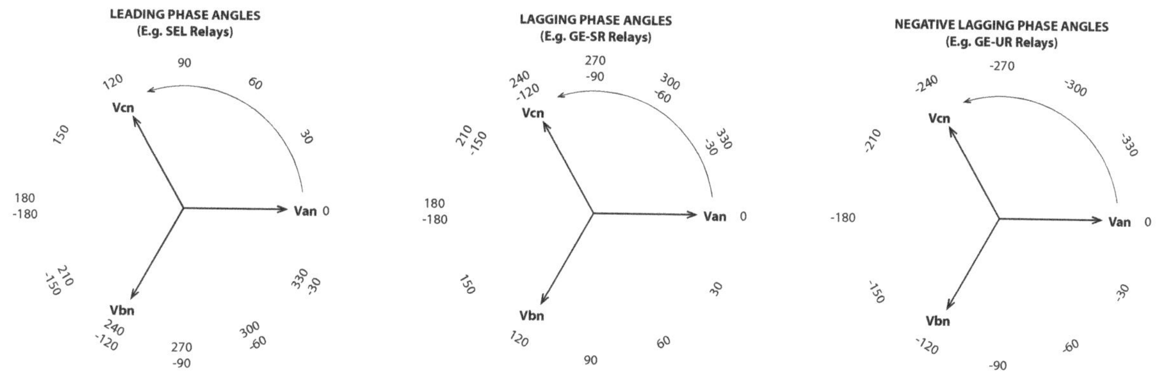

Figure 2-45: Phase Angle Relationships

Watt and VAR measurements can also help determine if the correct connections have been made. When three-phase, balanced current and voltages are applied, maximum watts and almost zero VARs should be measured. Rotate all three currents by 90° and maximum VARs and almost zero watts should be measured. Any connection problems will skew the watt and VAR values and should be corrected.

If the relay has a neutral input for voltage and/or current, apply a nominal value and compare metering values to the CT/PT ratios. A completed test sheet can look like the following:

METERING							
CURRENT (AMPS)							
SEC INJ INPUT	A PH	B PH	C PH	MFG (A)	% ERROR		
5.00	**2007**	**2007**	**2006**	2000	**0.35**	**0.35**	**0.30**
PHASE ANGLE	**31 LAG**	**150 LAG**	**270 LAG**				
SEC INJ INPUT	Pos Seq	Neg Seq	Zero Seq	MFG	% ERROR		
5.00	**2000**	**2001**	**2000**	2000	**0.00**	**0.05**	**0.00**
ROTATION	**ABC**	**ACB**	**A-G**				
VOLTAGE (VOLTS)							
SEC INJ INPUT	A PH	B PH	C PH	MFG (kV)	% ERROR		
120.00	**4.200**	**4.200**	**4.220**	4.200	**0.00**	**0.00**	**0.48**
PHASE ANGLE	**0 LAG**	**120 LAG**	**240 LAG**				
SEC INJ INPUT	Pos Seq (kV)	Neg Seq (kV)	Zero Seq (kV)	MFG (kV)	% ERROR		
120.00	**4.2000**	**4.2100**	**N/A**	4.2000	**0.00**	**0.24**	
ROTATION	**ABC**	**ACB**					
3-PHASE METERING							
FREQUENCY (Hz)				**POWER FACTOR**			
SEC INJ INPUT	3 PH (Hz)	MFG (Hz)	% ERROR	SEC INJ INPUT	3 PH	MFG	% ERROR
57.50	**57.50**	57.50	**0.00**	30 Degrees Lag	**0.866**	0.866	**0.00**
60.00	**60.00**	60.00	**0.00**	0 degrees	**1.000**	1.000	**0.00**
62.50	**62.50**	62.50	**0.00**	30 Degrees Lead	**-0.866**	-0.866	**0.00**
POWER (MW)				**VARS (MVAR)**			
SEC INJ INPUT	3 PH (MW)	MFG (MW)	% ERROR	SEC INJ INPUT	3 PH	MFG	% ERROR
1039.23	**14.566**	14.549	**0.12**	1039.23	**14.570**	14.549	**0.14**
-1039.23	**-14.556**	-14.549	**0.05**	-1039.23	**-14.552**	-14.549	**0.02**
1800 @ 90 Deg	**0.25**	0.00	**OK**	1800 @ 0 Deg	**0.360**	0.00	**OK**
COMMENTS:							
RESULTS ACCEPTABLE:		☑ YES		☐ NO		☐ SEE NOTES	

Figure 2-46: Example Metering Test Sheet

7. Apply the Test Plan

The end-to-end tests begin after all nodes have reported that their meter tests have been completed successfully. All nodes agree on a test case to run and load their respective test cases into the test-sets. An example test case for two nodes is shown in the following figures.

RLY-1	TEST-SET SECONDARY VALUES					
	PREFAULT		FAULT 1		POSTFAULT	
VA	66.4	0.0	28.9	-4.0	66.4	0.0
VB	66.4	-120.0	69.5	-123.0	66.4	-120.0
VC	66.4	120.0	69.0	123.0	66.4	120.0
IA	0.50	-10.0	37.99	-83.0	0.00	0.0
IB	0.50	-130.0	0.59	-89.0	0.00	0.0
IC	0.50	110.0	0.59	-79.0	0.00	0.0
Time (cycles)	60.00		10.00		60.0	

RLY-2	TEST-SET SECONDARY VALUES					
	PREFAULT		FAULT 1		POSTFAULT	
VA	66.4	0.0	60.1	-1.0	66.4	0.0
VB	66.4	-120.0	66.1	-119.0	66.4	-120.0
VC	66.4	120.0	66.2	119.0	66.4	120.0
IA	0.33	170.0	2.76	-80.0	0.00	0.0
IB	0.33	50.0	0.39	91.0	0.00	0.0
IC	0.33	-70.0	0.39	101.0	0.00	0.0
Time (cycles)	60.00		10.00		60.0	

Figure 2-47: Example Test Plan 2 with Raw Data

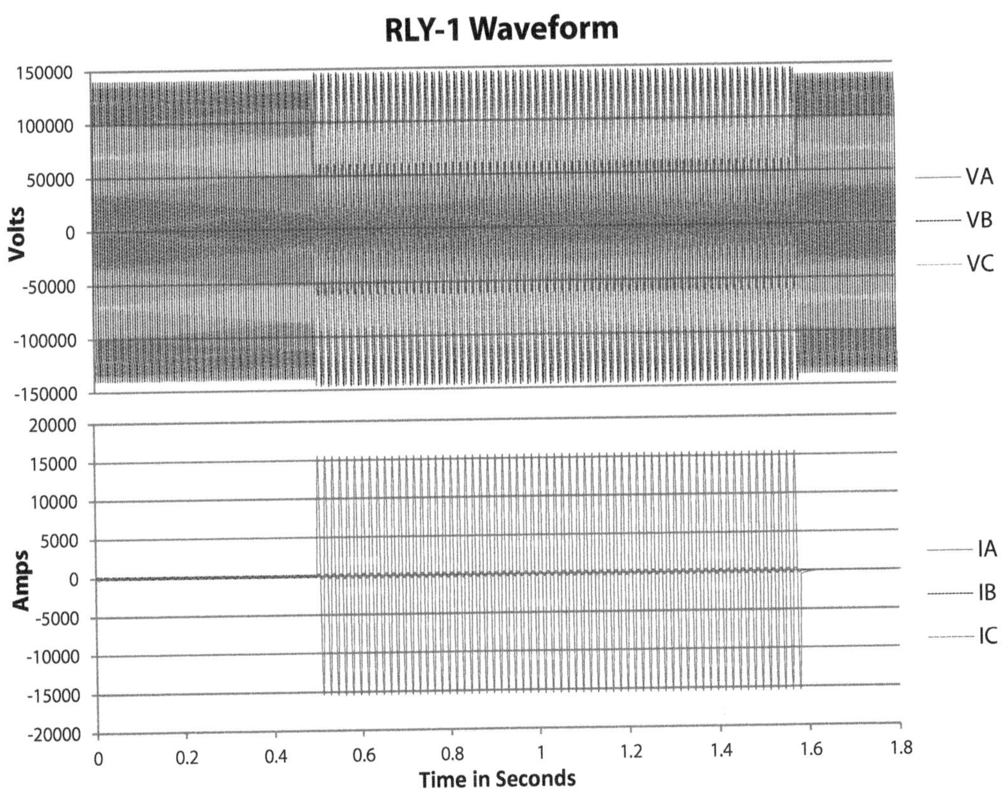

Figure 2-48: Example RLY-1 Waveform Test Plan 2

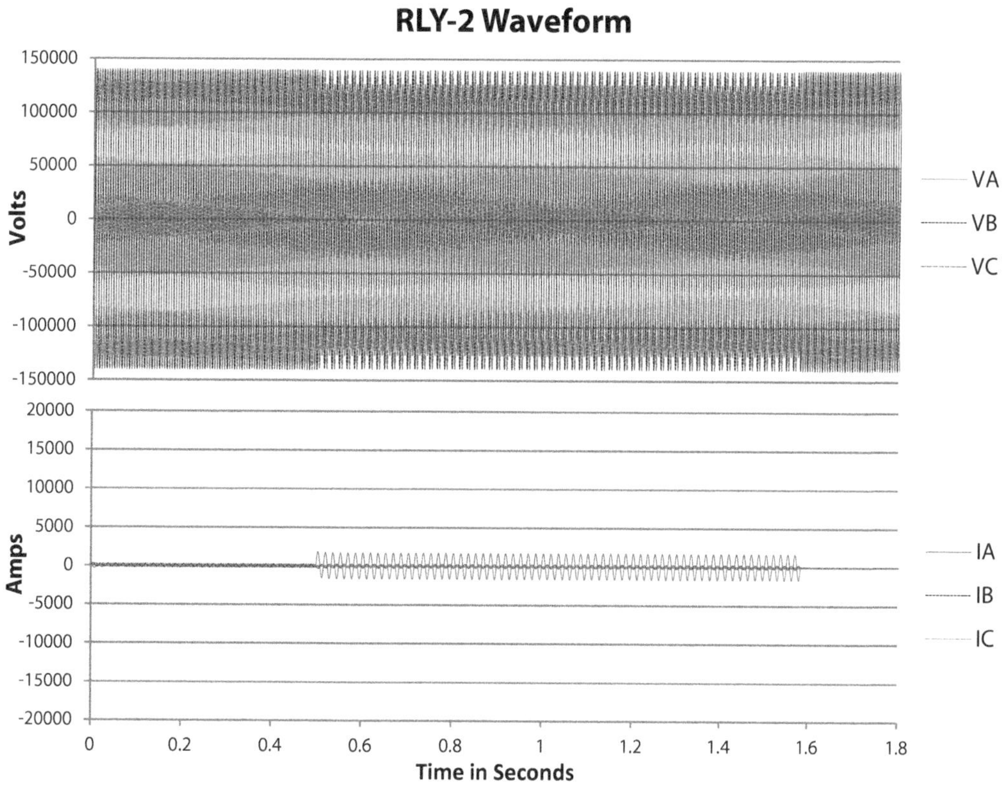

Figure 2-49: Example RLY-2 Waveform Test Plan 2

After the test cases are loaded into the test-sets, all sides should agree to a start time. Different test-sets have different methods to initiate a test. The test-set manufacturer(s) should be contacted to determine the correct method to initiate a test and whether some lead time should be added to ensure correct simulations when two different test-set models are used. This is particularly important when using Doble Laboratories test-sets because different firmware revisions could cause different time delays between the set time and actual starting time.

The test countdown should be initiated after all the start signals are synchronized for the same start time and all nodes report that they are ready. One side should call out the countdown and the other sides should verify they have the same countdown. Watch the test-set and relay metering, if possible, during the test to ensure the test-set started correctly and injected the correct values. If any nodes report a malfunction, the problem should be corrected and the test should be run again.

8. Evaluate the Results

After the test case has been injected into all nodes simultaneously, the relay targets at each node should be recorded and compared to the test case description to ensure the relays have responded correctly. The phase targets on the relay's front display (A, B, C, N, G) may not match the injected fault characteristic if you have programmed your test-set to stop when the relay operates. Some digital relays expect to use the time between the relay operation and actual breaker opening (up to 3 cycles) for targeting, but a test-set typically stops the applied voltage and current immediately after the relay trips. Stopping the currents and voltages too quickly hinders the relay's targeting algorithm and can cause incorrect phase targets. If you always want correct targets you should apply one of the following solutions depending on your test-set manufacturer's capabilities:

- Apply a breaker clearing timer that will automatically generate voltages and currents for a preset time after a stop signal is received.
- Use the breaker's 52a or 52b contact for a stop signal instead of the relay trip contact.
- Set a second state with identical settings that will be applied when the relay operates and then automatically stops after a preset delay.

Download all event and sequence-of-event records. Some engineers will review the event records from all relays to confirm the relays' reactions to the fault, and others will assume correct operation based on correct targeting and time delays for trips. There may be more than one oscillography file per trip depending on your relay settings.

(Hint: It's often a good idea to reset all fault recorder and sequence of event records inside the relay between tests to make sure that only the test in question is available to prevent confusion after several tests have been performed.)

If everything works correctly, all nodes can move on to the next test case and inject it into the relay.

A perfectly executed series of end-to-end tests is rare and there is often some troubleshooting involved. Here are some common problems that could cause an incorrect test result:

- What targets are displayed? If the SOTF target is displayed, the breaker was not closed during the test.
- Waveform Playback
  - Was the correct waveform loaded at all nodes?
- State Simulator Playback
  - Check the hard copy report of simulations to the data in the test-set
- Is the same Prefault duration applied at all nodes?
- Are the same phases faulted at all nodes?
- Are the phase angle references correct at all nodes?

- Did the playback start at the correct time at all nodes? (Look at relay event records)
- Were all AC channels recorded in the relay event records? (Test lead fell off, etc.)
- Were communication channels active during the test?
- Were the circuit breakers or circuit breaker simulators closed before the test?

9. Return the Protection System to Service

The protection system should be returned or placed into service after all test cases have been executed. Make sure that all event recorder, event records, and as-left relay settings have been downloaded and are available for off-site review before returning the relays into service. All event recorder, event records, min/max, and other history related data inside the relay should be erased to prevent confusion when troubleshooting faults after the relays have been placed in service. All test equipment should be disconnected from the relays and any wiring removed during the test should be replaced. The CT, PT, and input test switches should be closed first, and the relay outputs should be verified to be in their normal state before closing the trip test switches. Release the equipment to the switching authority after all test equipment is clear and the panels are free from interference.

Verify that the relay metering is correct after the equipment is energized to make sure all channels are connected properly.

10. Prepare the Report

The final report should document all of the test results, comments, and a final copy of the relay settings to allow the project manager to review the results and final settings. The following items should be included in every test report:

A) Cover Letter

The cover letter should describe the project, provide a brief history, and (most importantly) include a list of all comments during the test. This letter summarizes all of the test sheets and should be written with non-electrical personnel in mind. Ideally, this document could be reviewed years after the testing date with a clear understanding of what tests were performed and their results. Any comments should be clearly explained with a brief history of any actions performed and their status at the time of the letter. Organize comments in order of importance and by relay, or relay type if the same comment applies to multiple relays. An example comment is, "The current transformer ratio on Drawing A and the supplied relay settings did not match. The design engineer was contacted and the correct ratio of 600:5 was applied to the relay settings and confirmed in the field. No further action is required."

B) Test Sheet

The test sheet should clearly show all the test results, including a printout of event and sequence-of-event records for each test to show what tests were performed and the relays' responses. A digital copy of all test cases should also be included in the report to allow future maintenance personnel to replay the same tests into the relay and evaluate their response during maintenance intervals.

C) Final Settings

The final, as-left settings should be documented at the end of the test sheet. A digital copy should also be saved, and all relay settings for a project should be made available to the client or design engineer for review and their final documentation. Setting files should be in the relay's native software and in a universal format such as word processor or pdf file to allow the design engineer to make changes, if required, and allow anyone else to review the settings without special software.

Chapter 3

Common Protection Schemes

The following sections are intended to provide a basic understanding of the most common protection schemes tested via end-to-end testing. Distance protection settings can be generalized as a percentage of the line they are protecting for the most part, and it is important to understand the logic behind basic distance protection before we review the communication protection schemes.

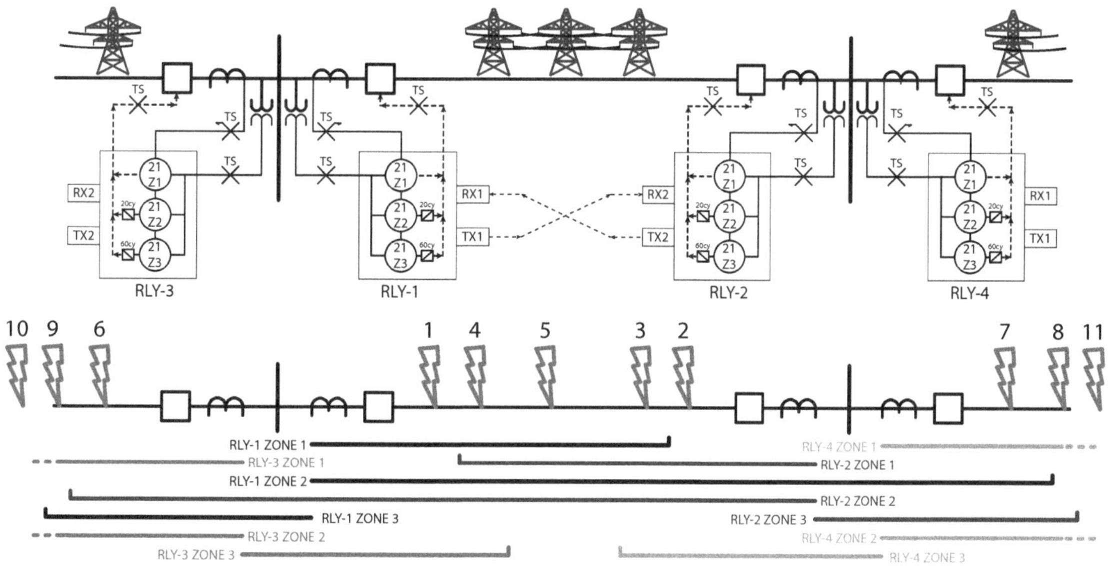

Figure 3-1: Typical Distance Protection Settings

Zone-1 protection is typically set at 80% of the transmission line with no intentional time delay. It is not set at 100% because:

- Most protection CTs have an accuracy rating of 10%.
- The line impedance used for protective relaying is typically a calculation based on the wire size and transmission line length, so there will be calculation errors.
- There are many other factors that can affect the actual line impedance such as manufacturers' tolerances, inexact distance measurements, splices, distance between conductors, etc., which make an exact calculation impossible.
- The actual fault will have unknown properties such as foreign material impedances, arc length, humidity, temperature, etc., which make exact fault distance calculations difficult.

Zone-1 reach settings between 70-90% are considered safe for instantaneous trips that will only detect faults on the transmission line.

Two protective relays with Zone-1 elements set at 80% of the line facing each other will provide 100% protection of the transmission line, including redundant protection on the inner 60% via the overlapping zones of protection. The average person would probably think this is good enough, but the utility industry is always concerned with reliability and stability, which requires 100% redundancy.

Zone-2 protection is usually set beyond (usually 120%) the transmission line to provide 100% redundancy for the protected transmission line. Zone-2 protection also provides backup protection for faults adjacent to the transmission line with a 20 cycle time delay to allow the adjacent relay to trip first. If the adjacent breaker does not open within 20 cycles, Zone-2 protection will operate to isolate the fault from the rest of the system. Zone-2 has a twofold benefit, redundant protection for the transmission line with a small time delay for faults located on the last 10-30% of the line not covered by Zone-1, and backup protection for external equipment.

Zone-3 protection in non-communication schemes can be applied to provide backup protection for external equipment. It can be applied with very large percentages in the forward direction with a long time delay (60 cycles) to minimize system disturbances in case of equipment failure. It can also be applied in the reverse direction with a similar time delay as backup protection for relays in the reverse direction.

The following figures depict the outcomes of a standard number of end-to-end tests assuming that the outside protective relays fail to operate for out-of-zone faults. Notice that the Zone-1 times should be zero but are shown as <3 cycles because no relay in the real world can operate instantaneously.

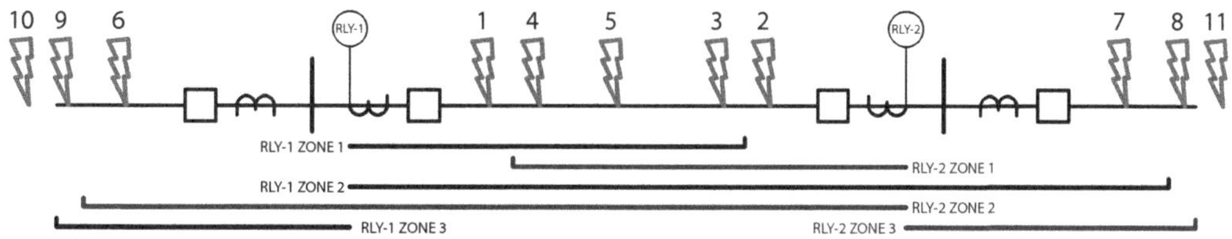

Figure 3-2: End-to-End Test Simulations

TEST	RLY-1	RLY-2
1	Trip Zone-1 in <3 cycles	Trip Zone-2 in 20 cycles
2	Trip Zone-2 in 20 cycles	Trip Zone-1 in <3 cycles
3	Trip Zone-1 in <3 cycles	Trip Zone-1 in <3 cycles
4	Trip Zone-1 in <3 cycles	Trip Zone-1 in <3 cycles
5	Trip Zone-1 in <3 cycles	Trip Zone-1 in <3 cycles
6	Trip Zone-3 in 60 cycles	Trip Zone-2 in 20 cycles
7	Trip Zone-2 in 20 cycles	Trip Zone-3 in 60 cycles
8	No trip	Trip Zone-3 in 60 cycles
9	Trip Zone-3 in 60 cycles	No trip
10	No trip	No trip
11	No trip	No trip

Figure 3-3: End-to-End Test Results with no Communication Scheme Applied

If you want to test your knowledge or see an animation of the standard distance protection scheme, go to the Valence Electrical Training Services blog forum at **http://relaytraining.com/distance-protection-animation/**.

One of the most common problems with end-to-end testing occurs when the breaker status is incorrectly applied. Some relays have a feature called Switch-On-To-Fault (SOTF) that is enabled to prevent unnecessary trip delays if the breaker is open, and then closes onto a fault condition. SOTF can use current to detect breaker status and, therefore, some Prefault current above the breaker energized settings should always be supplied to the relay before applying a fault. However, some relays use the 52a or 52b contacts to determine breaker status and will trip instantaneously if you apply current without changing the input breaker status. (When testing with breakers out of service and open, for example.) The element often uses Zone-2 and ground overcurrent to detect a fault and if you apply a fault inside Zone-2 without closing the breaker first, the relay will trip nearly instantaneously. Figure 3-4 depicts the end-to-end test results of SOTF. If you compared Figure 3-4 to Figure 3-3, you will see they are very similar, and a tester may be fooled into thinking that everything is working well until Test #6. Always make sure you know what the SOTF settings are, and you check for SOTF operation after every test.

TEST	RLY-1	RLY-2
1	Trip Zone-1 in <3 cycles	***Trip Zone-2 in <3 cycles***
2	***Trip Zone-2 in <3 cycles***	Trip Zone-1 in <3 cycles
3	Trip Zone-1 in <3 cycles	Trip Zone-1 in <3 cycles
4	Trip Zone-1 in <3 cycles	Trip Zone-1 in <3 cycles
5	Trip Zone-1 in <3 cycles	Trip Zone-1 in <3 cycles
6	Trip Zone-3 in 60 cycles	***Trip Zone-2 in <3 cycles***
7	***Trip Zone-2 in <3 cycles***	Trip Zone-3 in 60 cycles
8	No trip	Trip Zone-3 in 60 cycles
9	Trip Zone-3 in 60 cycles	No trip
10	No trip	No trip
11	No trip	No trip

Figure 3-4: End-to-End Test Results with SOTF

The most common communication protection schemes include:

1. Direct Transfer Trip (DTT) Scheme

The direct transfer trip scheme is the simplest of the communications schemes and allows a transfer-trip signal to be sent to all relays. If the correct local trip signal is detected on one relay, a transfer-trip signal is sent to all the other relays. End-to-end testing is not required for this communication scheme because the relays are sending commands instead of communicating with each other in real time. Therefore, a tester could create a transfer-trip condition at each end and verify that all other ends have opened as depicted in Figure 3-5.

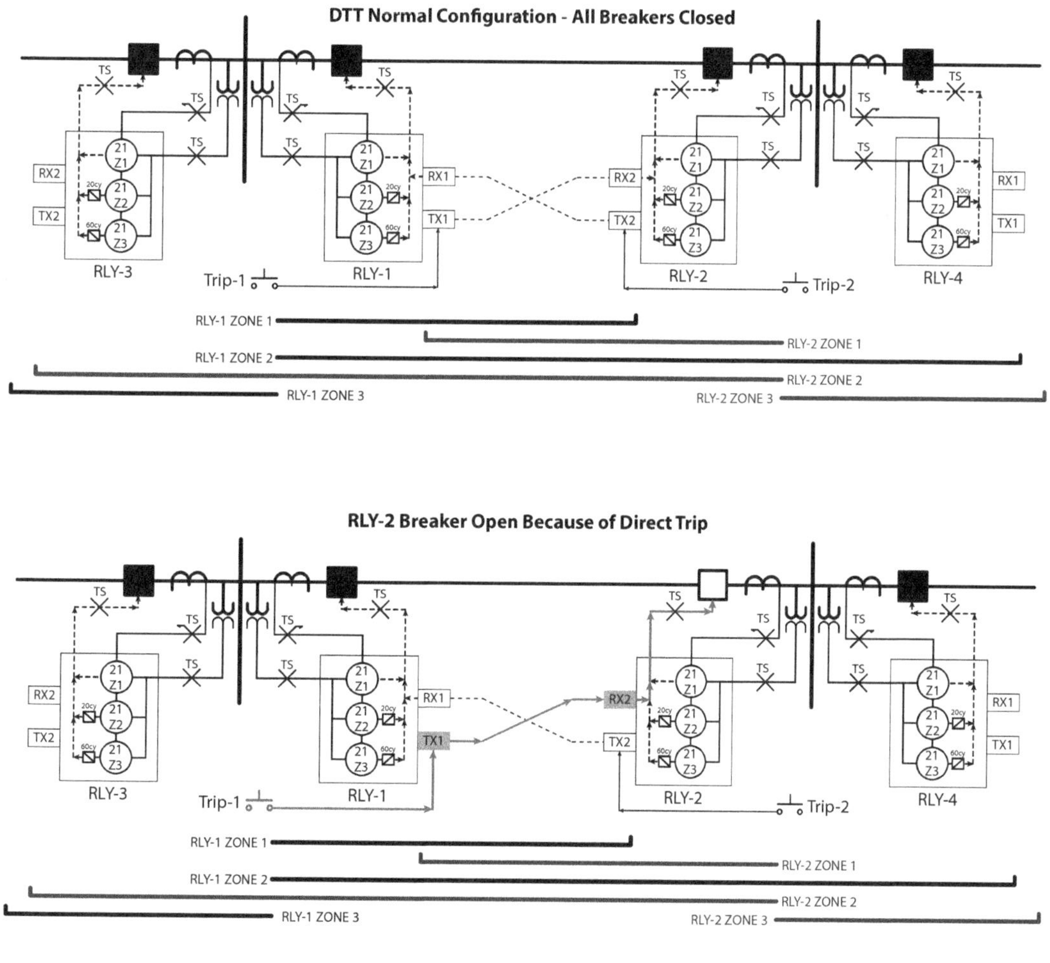

Figure 3-5: DTT Example

2. Direct Under-Reaching Transfer Trip (DUTT)

The direct under-reaching transfer trip scheme is very similar to the DTT scheme described above but it uses the Zone-1 protective element in each relay to send a DTT signal. Any relay that detects a Zone-1 fault will send a trip signal to all the other relays in the scheme. End-to-end testing is not required for this communication scheme because the relays are sending commands instead of communicating with each other in real time. Therefore, a tester could inject a Zone-1 condition at each end and verify that all other ends have opened as per the example in Figure 3-6.

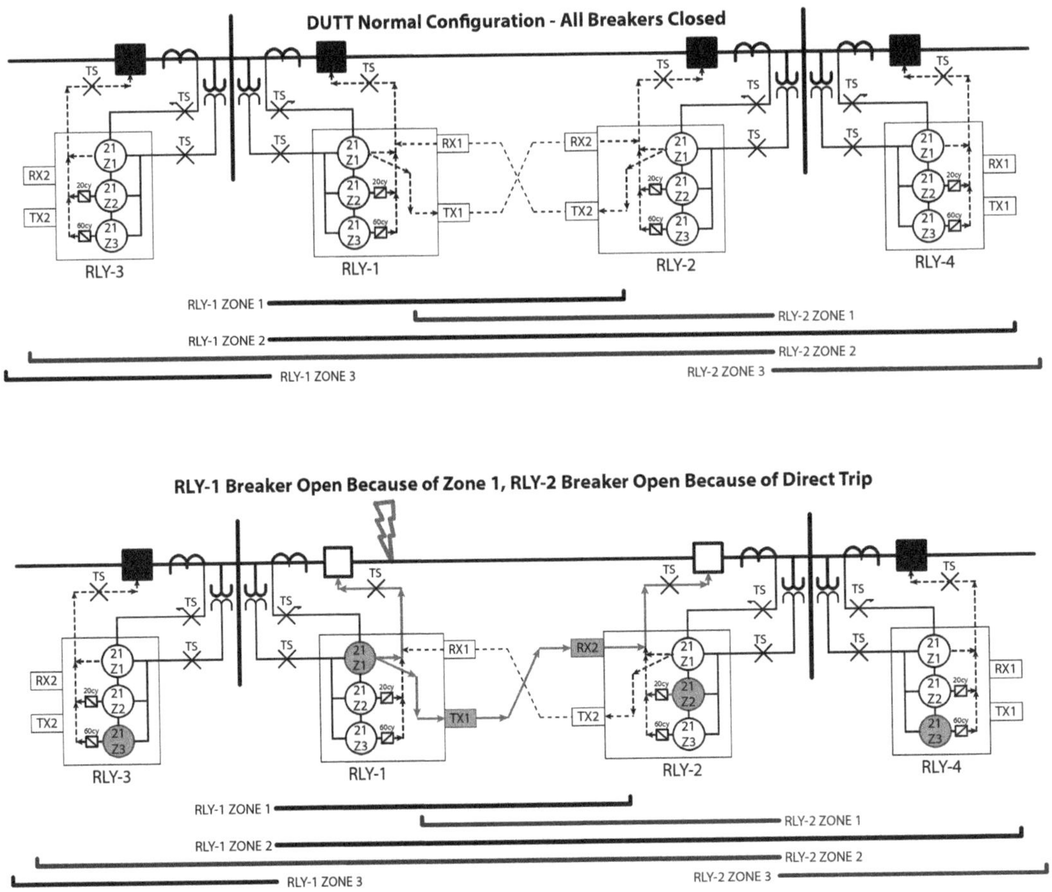

Figure 3-6: DUTT Example

If you want to test your knowledge or see an animation of the DUTT protection scheme, go to the Valence Electrical Training Services blog at **http://relaytraining.com/dutt-animation/**.

3. Permissive Over-Reaching Transfer Trip (POTT)

The permissive over-reaching transfer trip scheme uses Zone-2 elements from any two relays to determine if a fault has occurred on a transmission line. This scheme uses the standard Zone-1 and Zone-2 protection settings but will send a permissive trip to the other relays in the communication scheme if a Zone-2 pickup is detected. If two relays detect a Zone-2 fault pickup (Test Cases 1, 2, 3, 4 and 5 in Figure 3-7), the fault must be located inside the scheme's zone of protection because the two Zone-2 settings only overlap across the transmission line itself. If both relays do not detect a Zone-1 condition (Test Cases 1 and 2 in Figure 3-7), one relay will trip due to a Zone-1 pickup and the other relay will trip after a small time delay to prevent communication errors that can cause nuisance operations.

The following figure indicates the results of a POTT communication scheme operating correctly, assuming that the outside equipment does not operate for out-of-zone faults. Notice that the only two tests that are different from the standard scheme are Tests 1 and 2.

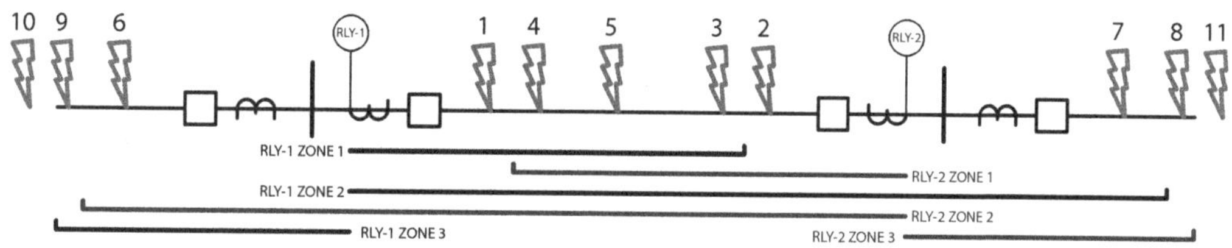

Figure 3-7: End-to-End Test Simulations

TEST	RLY-1	RLY-2
1	Trip Zone-1 in <3 cycles	***Trip Zone-2 in <6 cycles***
2	***Trip Zone-2 in <6 cycles***	Trip Zone-1 in <3 cycles
3	Trip Zone-1 in <3 cycles	Trip Zone-1 in <3 cycles
4	Trip Zone-1 in <3 cycles	Trip Zone-1 in <3 cycles
5	Trip Zone-1 in <3 cycles	Trip Zone-1 in <3 cycles
6	Trip Zone-3 in 60 cycles	Trip Zone-2 in 20 cycles
7	Trip Zone-2 in 20 cycles	Trip Zone-3 in 60 cycles
8	No trip	Trip Zone-3 in 60 cycles
9	Trip Zone-3 in 60 cycles	No trip
10	No trip	No trip
11	No trip	No trip

Figure 3-8: End-to-End Test Results with POTT Communication Scheme Applied

The following graphics depict a normal condition, a POTT trip, and a trip outside of the zone of protection. Always remember that Zone-2 will always have a time delay if the correct communication permissives are not received.

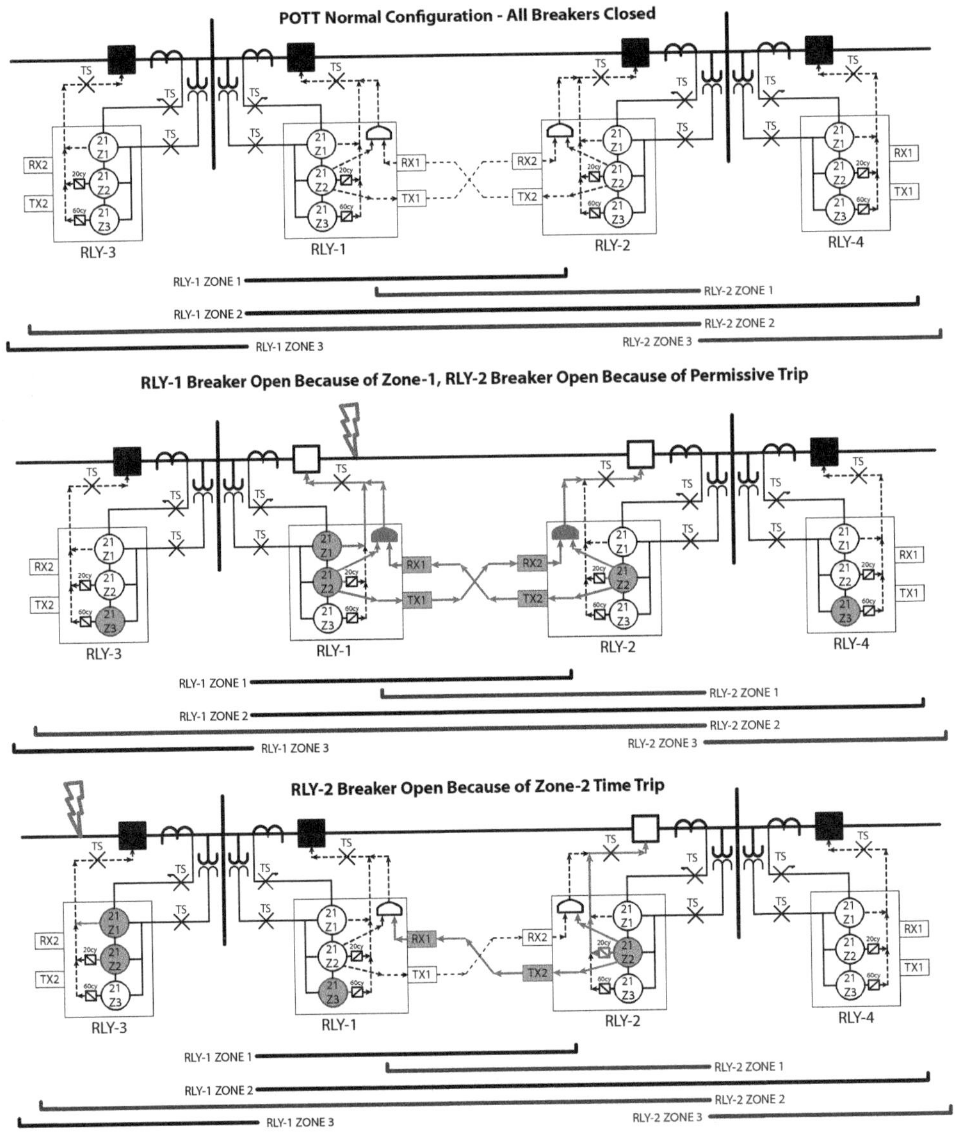

Figure 3-9: POTT Example

POTT schemes are the most common communication-schemes applied, but early applications of the POTT scheme discovered a weakness. POTT schemes can operate erratically during a sudden current-reversal because the scheme uses Zone-2 as the permissive signal, and Zone-2 over-reaches the transmission line into other zones of protection.

Sudden current-reversals can occur on installations such as parallel lines as shown in Figure 3-10. In the first example, all four breakers are closed and a fault occurs in Zone-1 of RLY-1. The current flows through each breaker as shown by the arrows. RLY-4 Zone-2 is picked up and sends a permissive signal to RLY-3. When Breaker-1 opens, the current suddenly reverses direction through RLY-3 and RLY-4, which starts a race. Will RLY-4 detect the sudden reversal first and stop sending the permissive trip to RLY-3 before RLY-3 detects a Zone-2 pickup? If not, RLY-3 will have a permissive signal from RLY-4 and detect a Zone-2 fault, which will cause a communication-assisted trip and de-energize the healthy feeder. You might think this is impossible, but you should always remember that nothing operates instantaneously in the real world and unexpected delays can often create a race between functions and create unintended operations.

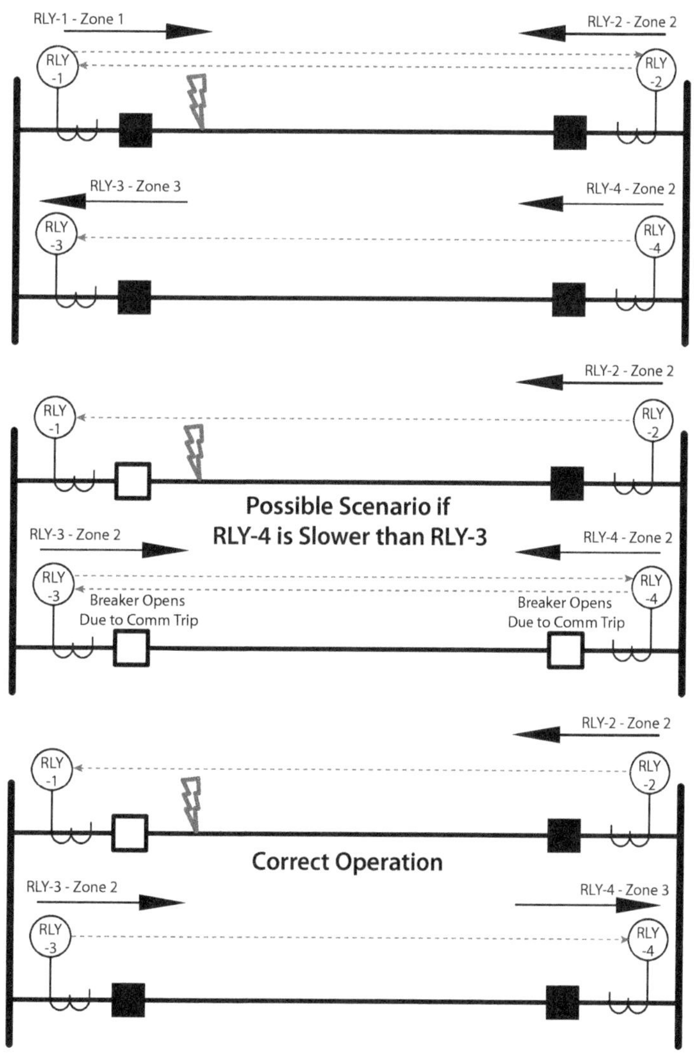

Figure 3-10: Current Reversal Example

Most POTT schemes have additional protection for preventing the phase reversal problems and can use Zone-3 to start a timer. If a relay detects a Zone-3, it cannot send a permissive signal to the other relay until a time delay has passed to prevent a race between elements.

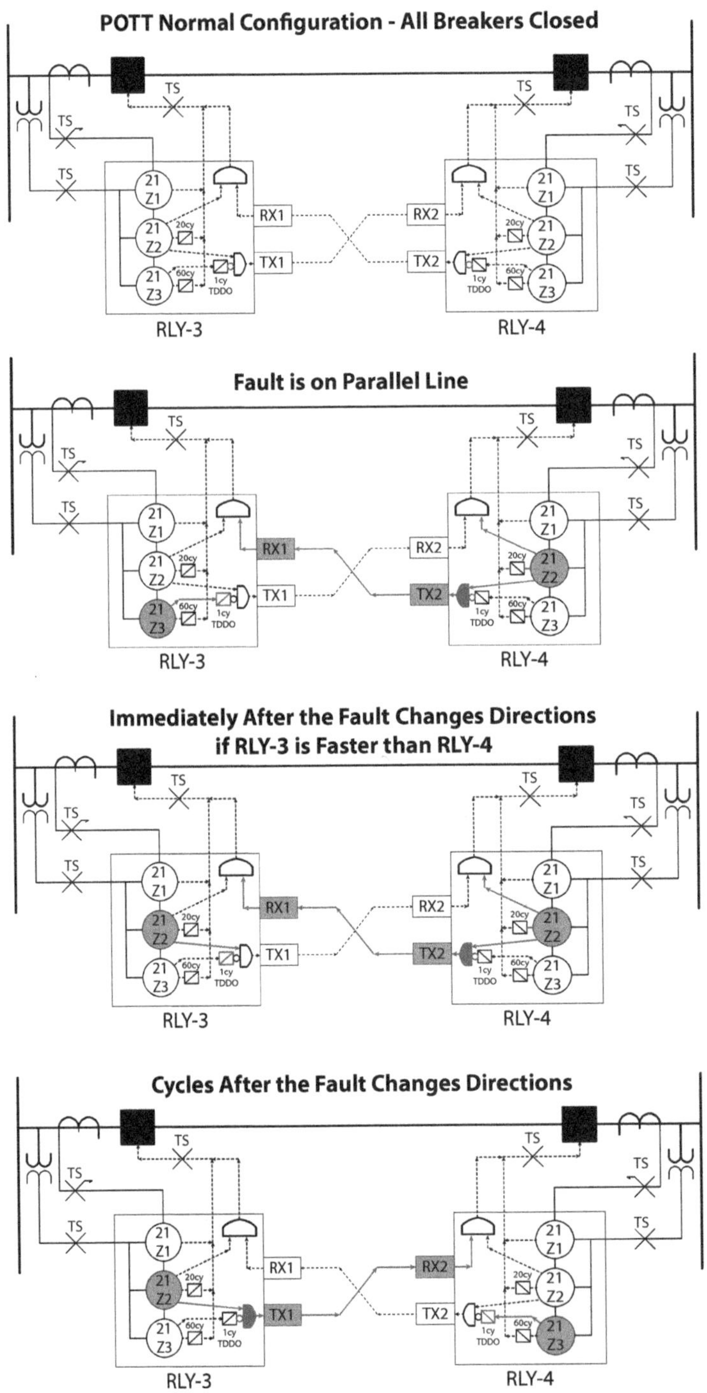

Figure 3-11: Modified POTT Scheme Example

Current-reversal blocking is tested by simulating a current-reversal as shown in the following figures.

RLY-1	Test Case #1							
	PREFAULT		FAULT 1		FAULT 2		Post Fault	
VA	67.06	0.0	66.08	-2.0	70.35	2.0	67.06	0.0
VB	67.06	-120.0	21.69	-124.0	4.59	-127.0	67.06	-120.0
VC	67.06	120.0	66.33	-122.0	68.28	117.0	67.06	120.0
IA	3.00	-20.0	3.03	-30.0	1.23	-20.0	0.00	0.0
IB	3.00	-140.0	17.96	-158.0	19.07	-22.0	0.00	0.0
IC	3.00	100.0	3.04	-29.0	1.24	-19.0	0.00	0.0
Time (cy)	30.00		10.00		10.00		30.0	

RLY-2	Test Case #2							
	PREFAULT		FAULT 1		FAULT 2		Post Fault	
VA	67.06	0.0	69.08	-1.0	66.06	-2.0	66.40	0.0
VB	67.06	-120.0	6.92	-125.0	19.52	-125.0	66.40	-120.0
VC	67.06	120.0	67.85	118.0	66.27	122.0	66.40	120.0
IA	3.00	160.0	3.03	150.0	1.23	15.0	0.00	0.0
IB	3.00	40.0	17.96	-22.0	19.07	158.0	0.00	0.0
IC	3.00	80.0	3.04	151.0	1.24	161.0	0.00	0.0
Time (cy)	30.00		10.00		10.00		30.0	

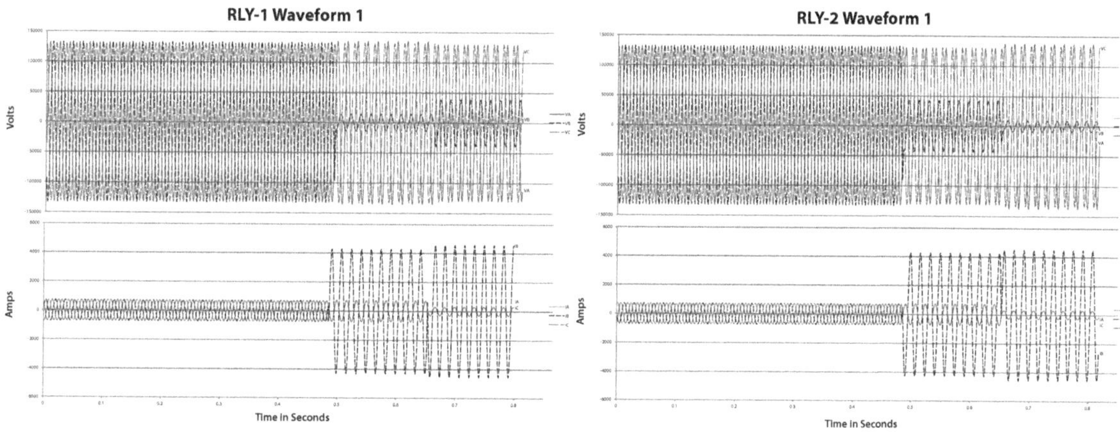

Figure 3-12: Current Reversal Test Plan

The POTT communication scheme speeds the tripping time for faults on the transmission line and can provide backup protection for faults outside of the zone.

If you want to test your knowledge or see an animation of the POTT protection scheme or see an animation of a sudden current reversal, go to the Valence Electrical Training Services blog at **http://relaytraining.com/pott-animation/**.

4. Directional Comparison Unblocking (DCUB)

The directional comparison unblocking scheme is fundamentally the same as the POTT scheme described previously. The relays communicate through a power line carrier channel that uses one phase of the power system as a communication channel. Communicating over the power line is less secure because the signals may not transmit across a faulted transmission line and power system disruptions can cause noise. Therefore, a guard or block signal is sent under normal conditions to indicate that communication channels are intact and no fault has been detected. If one relay detects a Zone-2 fault, the guard/block signal is dropped and a trip/unblock signal is sent. The other relay will trip almost immediately (after a small communications delay) if it detects the following:

- A local Zone-2 fault
- No guard signal immediately followed by a remote Zone-2 fault (The unblock signal)
- The lost guard signal and unblock signal must occur within a preset time delay as depicted by the timer in the figures. If this timer did not exist, any Zone-2 fault would trip nearly instantaneously if there was something wrong with the communication channel because a blocking signal would not be present.

The following figure indicates the results of a DCUB communication scheme operating correctly, assuming that the outside equipment does not operate for out-of-zone faults. Notice that the only two tests that are different from the standard scheme are Tests 1 and 2.

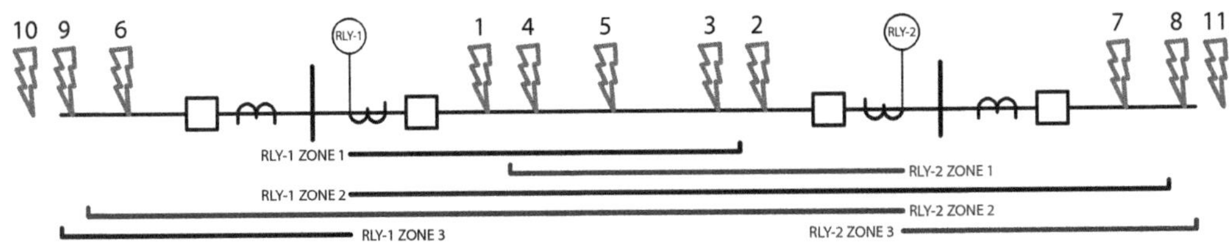

Figure 3-13: End-to-End Test Simulations

TEST	RLY-1	RLY-2
1	Trip Zone-1 in <3 cycles	***Trip Zone-2 in <6 cycles***
2	***Trip Zone-2 in <6 cycles***	Trip Zone-1 in <3 cycles
3	Trip Zone-1 in <3 cycles	Trip Zone-1 in <3 cycles
4	Trip Zone-1 in <3 cycles	Trip Zone-1 in <3 cycles
5	Trip Zone-1 in <3 cycles	Trip Zone-1 in <3 cycles
6	Trip Zone-3 in 60 cycles	Trip Zone-2 in 20 cycles
7	Trip Zone-2 in 20 cycles	Trip Zone-3 in 60 cycles
8	No trip	Trip Zone-3 in 60 cycles
9	Trip Zone-3 in 60 cycles	No trip
10	No trip	No trip
11	No trip	No trip

Figure 3-14: End-to-End Test Results with DCUB Communication Scheme Applied

The following graphics depict a normal condition, a DCUB trip, and a trip outside of the zone of protection. Always remember that Zone-2 will always have a time delay if the correct communication signals are not received.

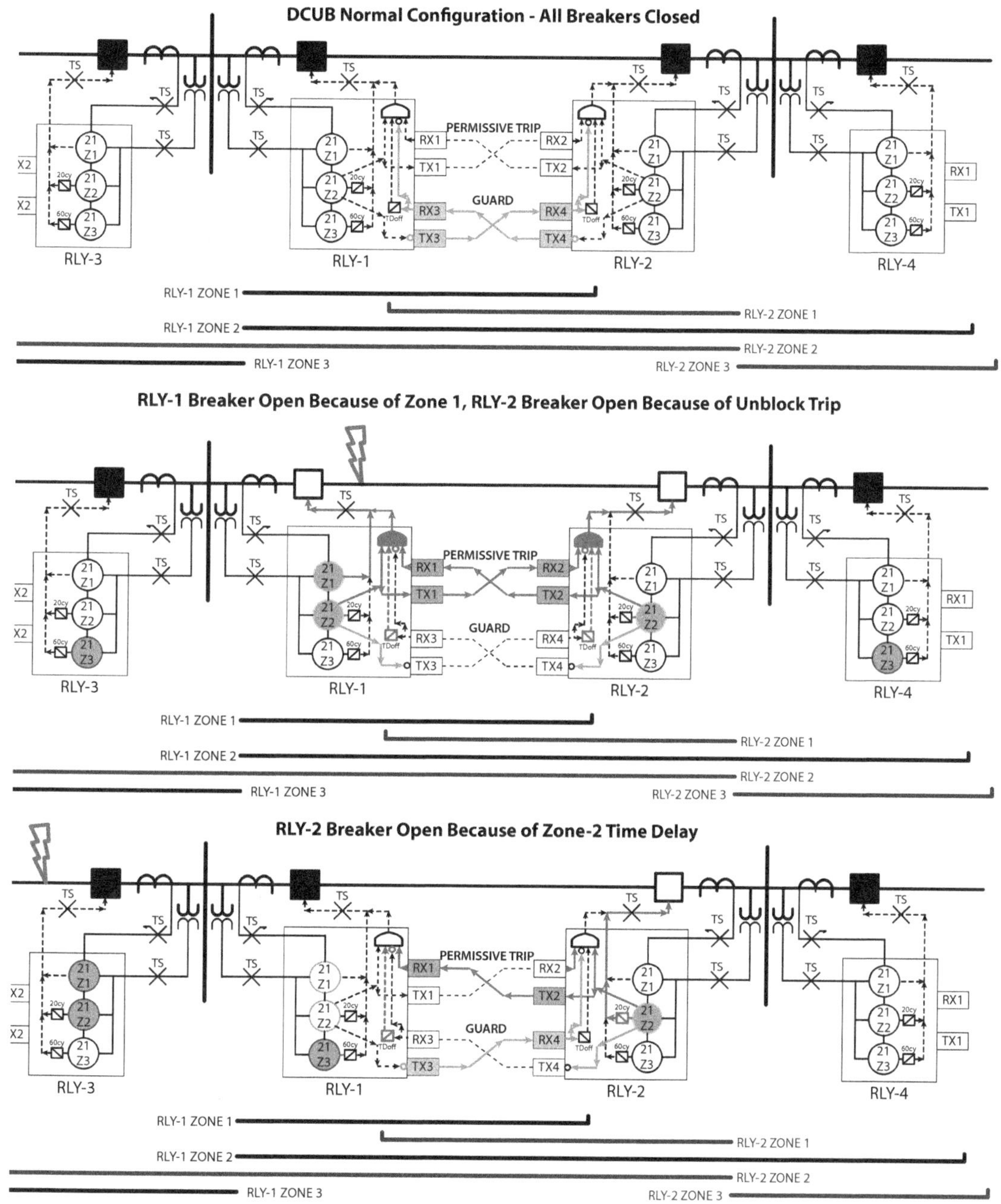

Figure 3-15: DCUB Example

If you want to test your knowledge or see an animation of the DCUB protection scheme, go to the Valence Electrical Training Services blog at **http://relaytraining.com/dcub-animation**.

5. Permissive Under-Reaching Transfer Trip (PUTT)

The permissive under-reaching transfer trip scheme uses the Zone-1 element from one relay and the Zone-2 element from a second relay to determine if a fault has occurred on the transmission line or outside the scheme's zone of protection. If a Zone-1 fault is detected by one relay (Test Cases 1, 2, 3, 4 and 5) and a Zone-2 fault pickup is detected by the other relay (Test cases 1 and 2), the fault must be located inside the scheme's zone of protection because the Zone-1 and opposite Zone-2 settings only overlap across the transmission line itself. The Zone-1 relay will trip instantaneously and the Zone-2 relay will trip after a small time delay is applied to prevent communication errors that can cause nuisance operations.

This scheme will not trip on sudden phase reversals because Zone-1 does not reach beyond the transmission line and will not operate for faults outside the zone. Zone-3 protection is not required for this scheme other than back-up protection, if desired.

The following figure indicates the results of a PUTT communication scheme operating correctly (assuming that the outside equipment does not operate for out-of-zone faults).

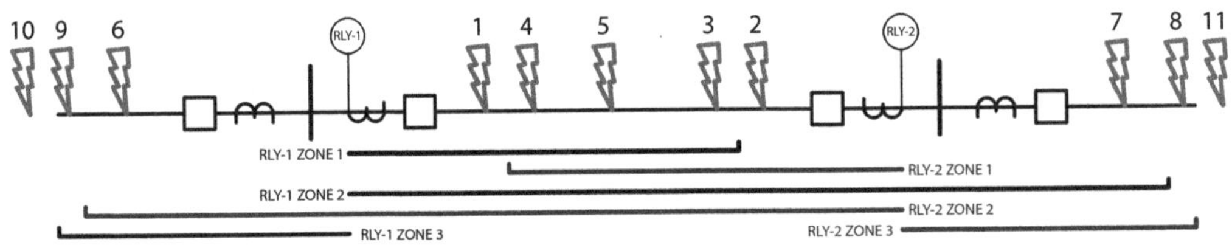

Figure 3-16: End-to-End Test Simulations

TEST	RLY-1	RLY-2
1	Trip Zone-1 in <3 cycles	Trip Zone-2 in <6 cycles
2	Trip Zone-2 in <6 cycles	Trip Zone-1 in <3 cycles
3	Trip Zone-1 in <3 cycles	Trip Zone-1 in <3 cycles
4	Trip Zone-1 in <3 cycles	Trip Zone-1 in <3 cycles
5	Trip Zone-1 in <3 cycles	Trip Zone-1 in <3 cycles
6	Trip Zone-3 in 60 cycles	Trip Zone-2 in 20 cycles
7	Trip Zone-2 in 20 cycles	Trip Zone-3 in 60 cycles
8	No trip	Trip Zone-3 in 60 cycles
9	Trip Zone-3 in 60 cycles	No trip
10	No trip	No trip
11	No trip	No trip

Figure 3-17: End-to-End Test Results with PUTT Communication Scheme Applied

The following graphics depict a normal condition, a PUTT trip, and a trip outside of the zone of protection. Always remember that Zone-2 will always have a time delay if the correct communication signals are not received.

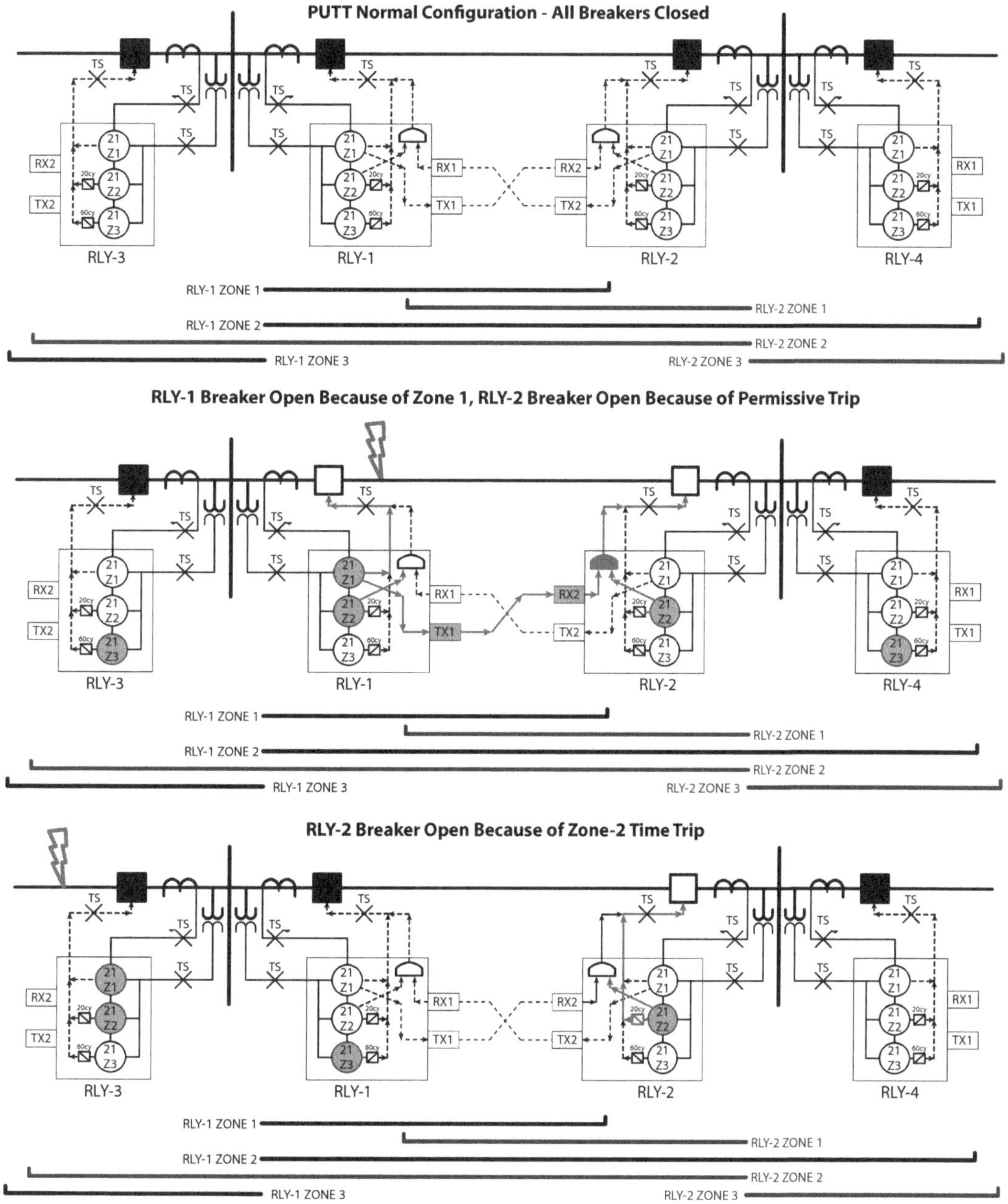

Figure 3-18: PUTT Example

If you want to test your knowledge or see an animation of the PUTT protection scheme, go to the Valence Electrical Training Services blog at **http://relaytraining.com/putt-animation.**

6. Directional Comparison Blocking (DCB)

The directional comparison blocking scheme is unique among the protection schemes described in this section because a blocking signal is used instead of a permissive signal. Zone-3 is set in the reverse direction and is typically set beyond the Zone-2 protection to provide a blocking signal to the other relays. The Zone-3 signal indicates the fault is not on the transmission line and sends that information to the other relays. The other relays will trip if they detect a Zone-2 pickup and do not receive the blocking signal within a small time delay.

This protection scheme is different from the previous ones because the Zone-2 timer typically operates within 6-10 cycles without a blocking signal from a remote relay. If a blocking signal is received, the relay will operate within the normal 15-30 cycle delay. This can cause confusion during normal relay testing because the Zone-2 time may have the standard 20 cycle setting but always trips faster because the remote relay is not sending a blocking signal.

If a Zone-2 fault is detected by one relay and the other relay does not detect a Zone-3 fault (Test Cases 1, 2, 3, 4 and 5), the first relay will assume that the fault is inside the designated zone of protection and will trip after a small time delay (< 6 cycles) is applied. The delay ensures that there is enough time to receive the blocking signal to prevent nuisance trips caused by communication delays.

If a Zone-2 fault is detected by one relay and the other relay detects a Zone-3 fault (Test Cases 6 and 7), a blocking signal will be sent to the first relay which will assume that the fault is outside the zone of protection and the normal Zone-2 timer will trip the breaker if the outside relays do not isolate the fault first.

The following figure indicates the results of a DCB communication scheme. (Assuming that the outside equipment does not operate for out-of-zone faults).

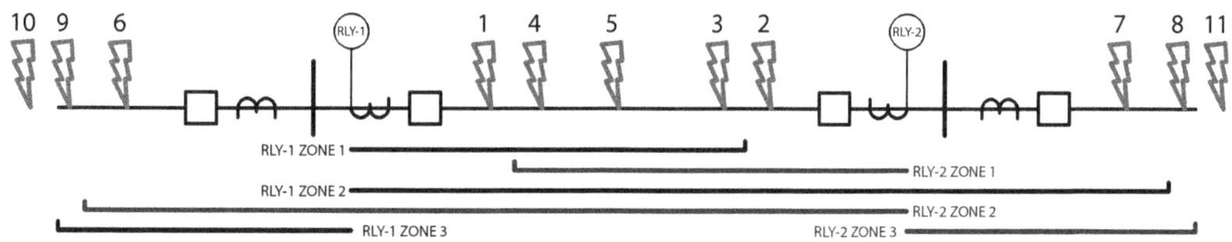

Figure 3-19: End-to-End Test Simulations

TEST	RLY-1	RLY-2
1	Trip Zone-1 in <3 cycles	***Trip Zone-2 in <6 cycles***
2	***Trip Zone-2 in <6 cycles***	Trip Zone-1 in <3 cycles
3	Trip Zone-1 in <3 cycles	Trip Zone-1 in <3 cycles
4	Trip Zone-1 in <3 cycles	Trip Zone-1 in <3 cycles
5	Trip Zone-1 in <3 cycles	Trip Zone-1 in <3 cycles
6	Trip Zone-3 in 60 cycles	Trip Zone-2 in 20 cycles
7	Trip Zone-2 in 20 cycles	Trip Zone-3 in 60 cycles
8	No trip	Trip Zone-3 in 60 cycles
9	Trip Zone-3 in 60 cycles	No trip
10	No trip	No trip
11	No trip	No trip

Figure 3-20: End-to-End Test Results with DCB Communication Scheme Applied

The following graphics depict a normal condition, a DCB trip, and a trip outside of the zone of protection.

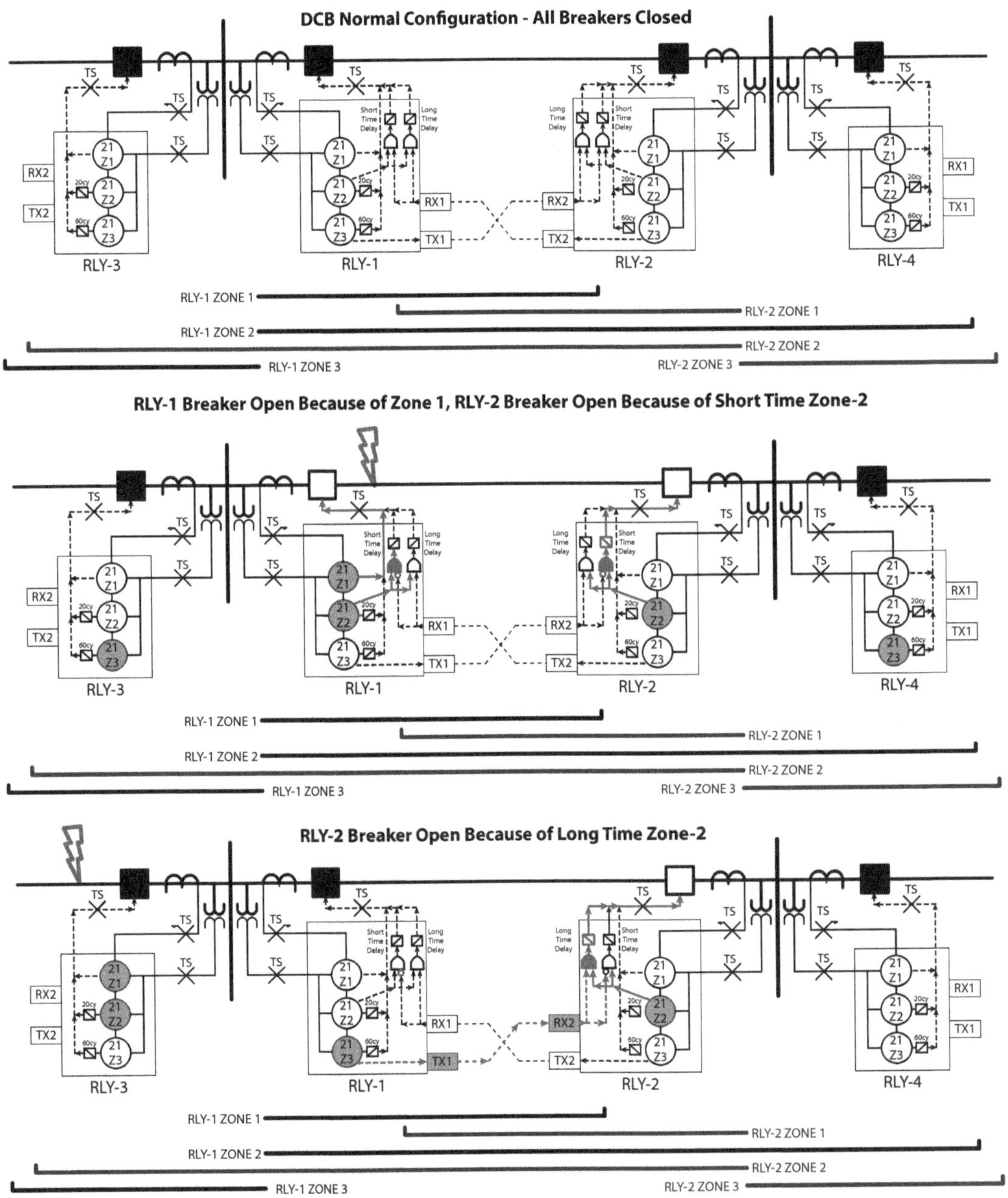

Figure 3-21: DCB Example

If you want to test your knowledge or see an animation of the DCB protection scheme, go to the Valence Electrical Training Services blog at **http://relaytraining.com/dcb-animation.**

8. Pilot Wire Protection

Pilot wire protection uses a relay located at each end of a transmission line that passes the three-phase CT secondary currents through filters and converts them into a signal that can be transmitted through a two-conductor pilot wire. The voltage or current signal passing through the pilot wire is connected to restraint and operating coils which operate as simple differential relays. The relays measure and compare the current entering the zone of protection to the current leaving the zone of protection. The difference between signals is the differential current. If the ratio between the differential current and restraint current exceeds the relays' setpoints, the relays will trip.

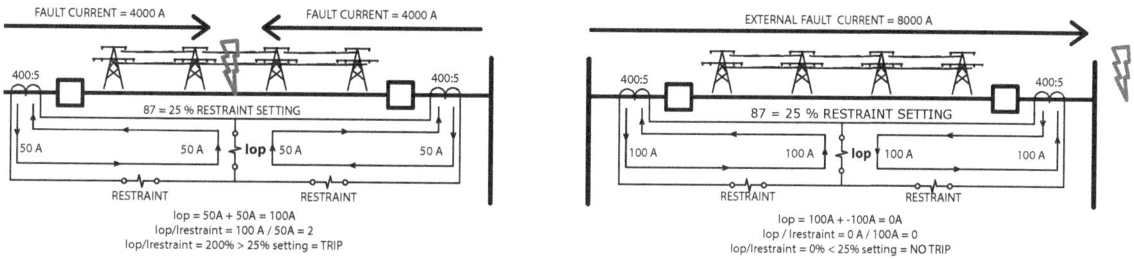

Figure 3-22: Simplified Pilot Wiring Operation

Faults that occur between the two relays will cause a trip and faults outside the zone will not trip as per the following chart.

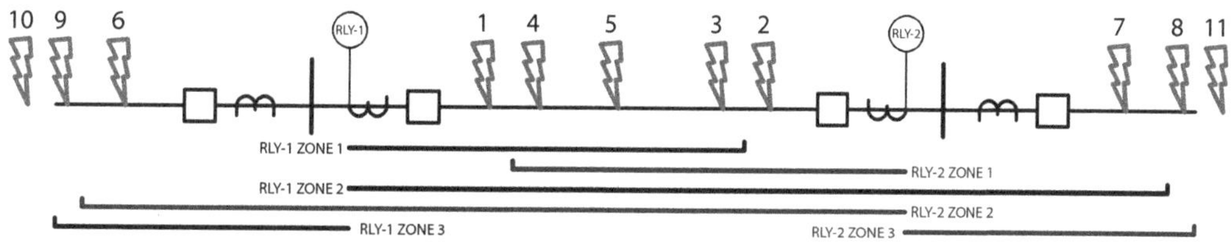

Figure 3-23: End-to-End Test Simulations

TEST	RLY-1	RLY-2
1	Trip 87 in <3 cycles	Trip 87 in <3 cycles
2	Trip 87 in <3 cycles	Trip 87 in <3 cycles
3	Trip 87 in <3 cycles	Trip 87 in <3 cycles
4	Trip 87 in <3 cycles	Trip 87 in <3 cycles
5	Trip 87 in <3 cycles	Trip 87 in <3 cycles
6-11	No trip	No trip

Figure 3-24: End-to-End Test Results with Pilot Wire Scheme Applied

9. Phase/Charge Comparison Protection

Phase/charge comparison protection uses a simple overcurrent device to initiate a trip that can be blocked if the relays detect that the fault is external to the zone of protection. If the measured current is greater than a fault-detection setpoint, each relay sends a blocking signal to the other relay during every positive half-cycle of the waveform and sends a trip signal every negative half cycle. A fault occurring on the transmission line with dual sources will cause both relays to send a trip signal on the negative half-cycle with no blocking signal and the relays will trip.

If only one side has a source, there will be no current on the opposite side to send a blocking signal and the relays will trip.

If the fault occurs outside the zone, the current phasors will be in opposition and the relays will not trip because a blocking signal appears to negate every trip signal as shown in the following figure.

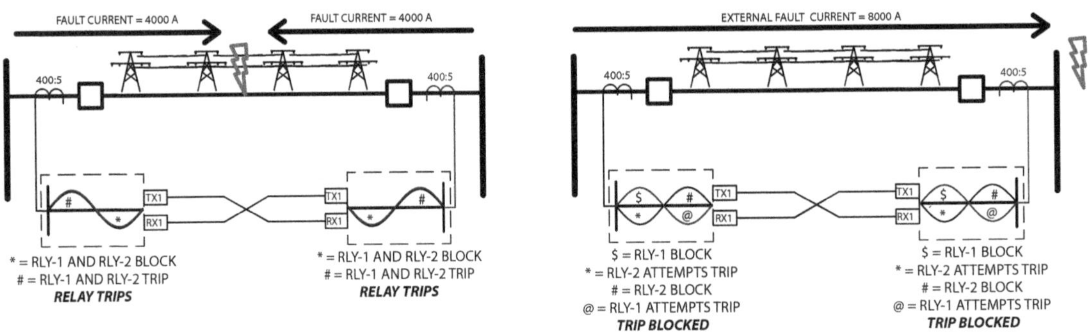

Figure 3-25: Phase/Charge Comparison Example of Operation

Faults that occur between the two relays will cause a trip and faults outside the zone will not trip as per the following chart.

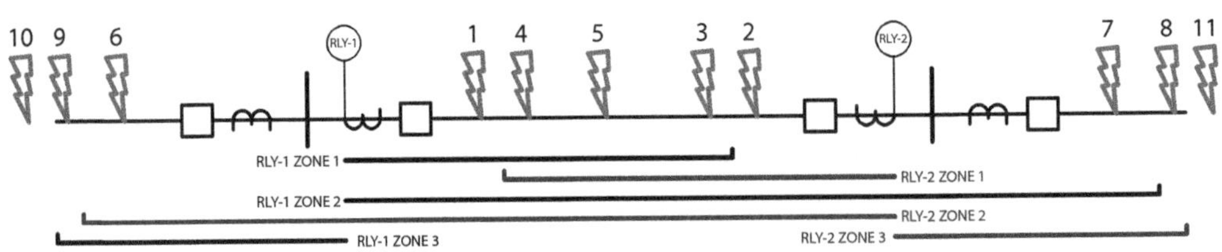

Figure 3-26: End-to-End Test Simulations

TEST	RLY-1	RLY-2
1	Trip 87 in <3 cycles	Trip 87 in <3 cycles
2	Trip 87 in <3 cycles	Trip 87 in <3 cycles
3	Trip 87 in <3 cycles	Trip 87 in <3 cycles
4	Trip 87 in <3 cycles	Trip 87 in <3 cycles
5	Trip 87 in <3 cycles	Trip 87 in <3 cycles
6-11	No trip	No trip

Figure 3-27: End-to-End Test Results with Phase/Charge Comparison Scheme Applied

10. Line Differential

Line differential protection has up to three relays connected at the ends of a transmission line that compare the magnitudes and phasors for all phases as well as the zero-sequence, negative-sequence, and residual ground to determine if a fault occurs on the transmission line or outside the zone of protection. These relays require a high-speed and secure communication medium such as fiber optics in order to transfer all of the collected analog data between all of the relays so each relay can perform its own calculations and provide high-speed tripping.

A) Standard Line Differential Relays

Traditional line current differential relays measure the local current entering the relay and send that information to all the other relays in the scheme. Each relay calculates the true secondary current flowing in and out of the transmission line by applying the local and remote CT ratio settings to the measured local and remote currents to compare apples to apples. The relay then acts as a standard differential relay and calculates the Operate and Restraint currents for each phase using the manufacturer's version of the following formulas.

$I_{Operate} = I_{Local} - I_{Remote}$	$I_{Restraint} = \frac{I_{Local} + I_{Remote}}{2}$	$\text{Slope (\%)} = 100 \times \frac{I_{Operate}}{I_{Restraint}}$

If the calculated slope is greater than the slope setting, the relay will trip. The following simplified examples demonstrate how the line differential relays work on a per-phase basis.

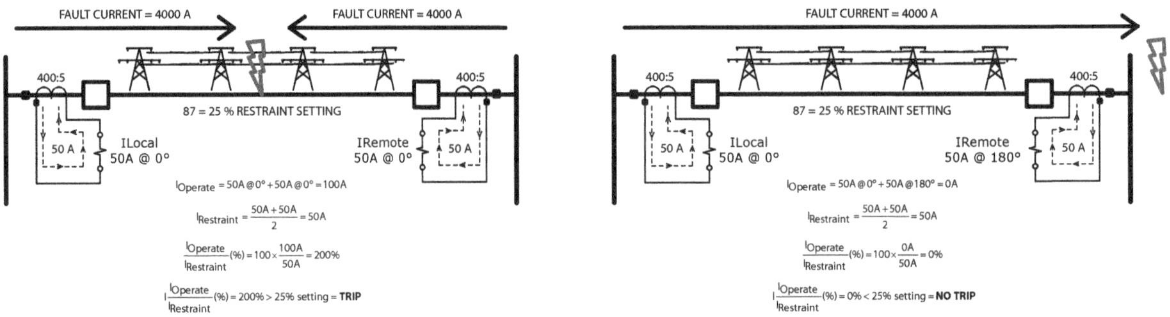

Figure 3-28: Standard Line Differential Example Calculations

B) The Alpha Plane

Schweitzer Engineering Laboratory's line differential relays use a characteristic called the alpha plane. The alpha plane graphs the vectoral ratio of remote current to local current calculated for each phase, as well as the negative and zero-sequence currents. The calculated values are then plotted on a graph of the alpha plane characteristic. Calculated values plotted inside the shaded area are in the restrained region and the relays will not trip. Values outside the shaded area are in the trip region and the relays will trip.

The following figures depict a Phase-Neutral fault on the transmission line including the fault parameters, the calculations, and how they would be plotted on the alpha plane. Notice that the following currents are inside the shaded region and will not cause a trip:

- PrefaultIA alpha
- PrefaultIB alpha
- PrefaultIC alpha
- FaultIB alpha
- FaultIC alpha

The following values fall outside the zone of protection and will cause the relays to trip:

- FaultIA alpha
- FaultI0 alpha
- FaultI2 alpha

There were no zero and negative-sequence quantities in Prefault, therefore those values were not plotted on the graph.

RLY-1	TEST-SET SECONDARY VALUES					
	PREFAULT		FAULT 1		POSTFAULT	
VA	66.4	0.0	28.9	-4.0	66.4	0.0
VB	66.4	-120.0	69.5	-123.0	66.4	-120.0
VC	66.4	120.0	69.0	123.0	66.4	120.0
IA	0.50	-10.0	37.99	-83.0	0.00	0.0
IB	0.50	-130.0	0.59	-89.0	0.00	0.0
IC	0.50	110.0	0.59	-79.0	0.00	0.0
Time (cycles)	60.00		10.00		60.0	

RLY-2	TEST-SET SECONDARY VALUES					
	PREFAULT		FAULT 1		POSTFAULT	
VA	66.4	0.0	60.1	-1.0	66.4	0.0
VB	66.4	-120.0	66.1	-119.0	66.4	-120.0
VC	66.4	120.0	66.2	119.0	66.4	120.0
IA	0.33	170.0	2.76	-80.0	0.00	0.0
IB	0.33	50.0	0.39	91.0	0.00	0.0
IC	0.33	-70.0	0.39	101.0	0.00	0.0
Time (cycles)	60.00		10.00		60.0	

Figure 3-29: Test Case #1 Values

PREFAULT	FAULT 1
$PrefaultIA_{alpha} = \frac{I_{RLY-2}}{I_{RLY-1}} = \frac{0.33@170^\circ}{0.50@-10^\circ} = 0.67@180^\circ$	$FaultIA_{alpha} = \frac{I_{RLY-2}}{I_{RLY-1}} = \frac{2.76@-80^\circ}{37.99@-83^\circ} = 0.07@3^\circ$
$PrefaultIB_{alpha} = \frac{I_{RLY-2}}{I_{RLY-1}} = \frac{0.33@50^\circ}{0.33@-130^\circ} = 0.67@180^\circ$	$FaultIB_{alpha} = \frac{I_{RLY-2}}{I_{RLY-1}} = \frac{0.39@91^\circ}{0.59@-89^\circ} = 0.66@180^\circ$
$PrefaultIC_{alpha} = \frac{I_{RLY-2}}{I_{RLY-1}} = \frac{0.33@-70^\circ}{0.33@110^\circ} = 0.67@180^\circ$	$FaultIC_{alpha} = \frac{I_{RLY-2}}{I_{RLY-1}} = \frac{0.39@101^\circ}{0.59@-79^\circ} = 0.66@180^\circ$
$PrefaultI0_{alpha} = \frac{I_{RLY-2}}{I_{RLY-1}} = \frac{0.00@0^\circ}{0.00@0^\circ} = 0.00@0^\circ$	$FaultI0_{alpha} = \frac{I_{RLY-2}}{I_{RLY-1}} = \frac{0.66@-78.4^\circ}{13.06@-83.0^\circ} = 0.05@4.6^\circ$
$PrefaultI2_{alpha} = \frac{I_{RLY-2}}{I_{RLY-1}} = \frac{0.00@0^\circ}{0.00@0^\circ} = 0.00@0^\circ$	$FaultI2_{alpha} = \frac{I_{RLY-2}}{I_{RLY-1}} = \frac{1.07@-80.6^\circ}{12.44@-83.0^\circ} = 0.09@2.4^\circ$

Figure 3-30: Test Case #1 Alpha Plane Calculations

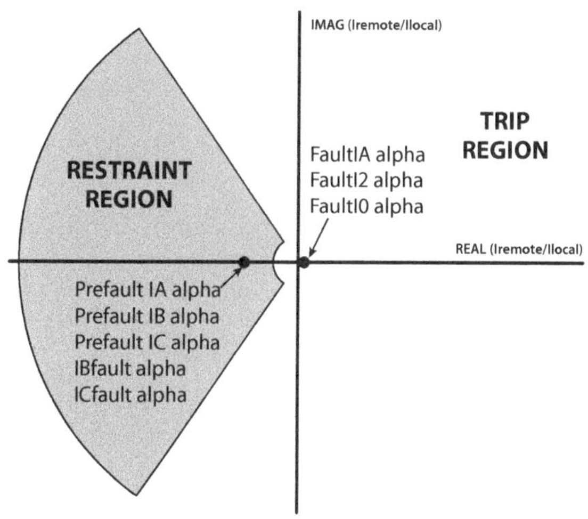

Figure 3-31: Test Case #1 Alpha Plane Graphed Calculations

The following figures depict a 3-Phase Fault on the transmission line including the fault parameters, the calculations, and how they would be plotted on the alpha plane. Notice that the following currents are inside the shaded region and will not cause a trip:

- PrefaultIA alpha
- PrefaultIB alpha
- PrefaultIC alpha

The following values fall outside the zone of protection and will cause the relays to trip:

- FaultIA alpha
- FaultIB alpha
- FaultIC alpha

There were no zero or negative-sequence quantities because the fault and prefault conditions are three-phase balanced and, therefore those values were not plotted on the graph.

RLY-1	TEST-SET SECONDARY VALUES					
	PREFAULT		FAULT 1		POSTFAULT	
VA	66.4	0.0	55.5	-1.0	66.4	0.0
VB	66.4	-120.0	55.5	-121.0	66.4	-120.0
VC	66.4	120.0	55.5	119.0	66.4	120.0
IA	0.50	-10.0	11.13	-81.0	0.0	0.0
IB	0.50	-130.0	11.13	159.0	0.0	0.0
IC	0.50	110.0	11.13	39.0	0.0	0.0
Time (cycles)	60.00		10.00		60.0	

RLY-2	TEST-SET SECONDARY VALUES					
	PREFAULT		FAULT 1		POSTFAULT	
VA	66.4	0.0	59.6	-1.0	66.4	0.0
VB	66.4	-120.0	59.6	-121.0	66.4	-120.0
VC	66.4	120.0	59.6	119.0	66.4	120.0
IA	0.33	170.0	7.97	-81.0	0.0	0.0
IB	0.33	50.0	7.97	159.0	0.0	0.0
IC	0.33	-70.0	7.97	39.0	0.0	0.0
Time (cycles)	60.00		10.00		60.0	

Figure 3-32: Test Case #2 Values

PREFAULT	FAULT 1
$PrefaultIA_{alpha} = \frac{I_{RLY-2}}{I_{RLY-1}} = \frac{0.33@170^{\circ}}{0.50@-10^{\circ}} = 0.67@180^{\circ}$	$FaultIA_{alpha} = \frac{I_{RLY-2}}{I_{RLY-1}} = \frac{7.97@-81^{\circ}}{11.13@-81^{\circ}} = 0.72@0^{\circ}$
$PrefaultIB_{alpha} = \frac{I_{RLY-2}}{I_{RLY-1}} = \frac{0.33@50^{\circ}}{0.33@-130^{\circ}} = 0.67@180^{\circ}$	$FaultIB_{alpha} = \frac{I_{RLY-2}}{I_{RLY-1}} = \frac{7.97@159^{\circ}}{11.13@159^{\circ}} = 0.72@0^{\circ}$
$PrefaultIC_{alpha} = \frac{I_{RLY-2}}{I_{RLY-1}} = \frac{0.33@-70^{\circ}}{0.33@110^{\circ}} = 0.67@180^{\circ}$	$FaultIC_{alpha} = \frac{I_{RLY-2}}{I_{RLY-1}} = \frac{7.97@39^{\circ}}{11.13@39^{\circ}} = 0.72@0^{\circ}$
$PrefaultI0_{alpha} = \frac{I_{RLY-2}}{I_{RLY-1}} = \frac{0.00@0^{\circ}}{0.00@0^{\circ}} = 0.00@0^{\circ}$	$FaultI0_{alpha} = \frac{I_{RLY-2}}{I_{RLY-1}} = \frac{0.00@0^{\circ}}{0.00@0^{\circ}} = 0.00@0^{\circ}$
$PrefaultI2_{alpha} = \frac{I_{RLY-2}}{I_{RLY-1}} = \frac{0.00@0^{\circ}}{0.00@0^{\circ}} = 0.00@0^{\circ}$	$FaultI2_{alpha} = \frac{I_{RLY-2}}{I_{RLY-1}} = \frac{0.00@0^{\circ}}{0.00@0^{\circ}} = 0.00@0^{\circ}$

Figure 3-33: Test Case #2 Alpha Plane Calculations

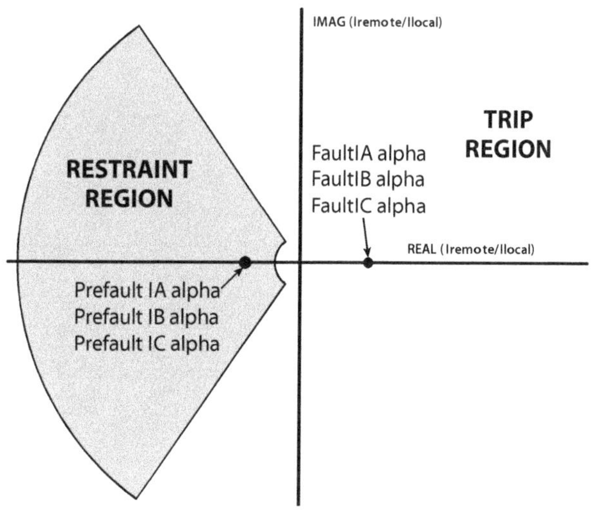

Figure 3-34: Test Case #2 Alpha Plane Graphed Calculations

The following figures depict a Phase-Phase fault on a parallel line with a phase-reversal and include the fault parameters, the calculations, and how they would be plotted on the alpha plane. Notice that all of the currents are inside the shaded region and will not cause a trip because the fault is not located on the line.

RLY-1	Test Case #1							
	PREFAULT		FAULT 1		FAULT 2		Post Fault	
VA	66.4	0.0	67.3	0.0	67.3	0.0	66.4	0.0
VB	66.4	-120.0	33.7	180.0	52.5	-130.0	66.4	-120.0
VC	66.4	120.0	33.7	180.0	52.3	129.0	66.4	120.0
IA	0.50	-10.0	0.06	-85.0	0.06	-85.0	0.0	-10.0
IB	0.50	-130.0	4.65	9.0	3.66	-171.0	0.0	-130.0
IC	0.50	110.0	4.65	-172.0	3.66	10.0	0.0	110.0
Time (cy)	60.00		10.00		10.00		60.00	

RLY-2	Test Case #2							
	PREFAULT		FAULT 1		FAULT 2		Post Fault	
VA	66.4	0.0	66.7	0.0	66.7	0.0	66.4	0.0
VB	66.4	-120.0	57.5	-126.0	34.1	-173.0	66.4	-120.0
VC	66.4	120.0	56.7	125.0	33.1	173.0	66.4	120.0
IA	0.33	170.0	0.04	95.0	0.04	95.0	0.0	170.0
IB	0.33	50.0	3.10	-171.0	2.44	9.0	0.0	50.0
IC	0.33	-70.0	3.10	8.0	2.44	-170.0	0.0	-70.0
Time (cy)	60.00		10.00		10.00		60.00	

Figure 3-35: Test Case #2 Values

$$\text{PrefaultIA}_{\text{alpha}} = \frac{I_{RLY-2}}{I_{RLY-1}} = \frac{0.33@170^\circ}{0.50@-10^\circ} = 0.67@180^\circ$$

$$\text{PrefaultIB}_{\text{alpha}} = \frac{I_{RLY-2}}{I_{RLY-1}} = \frac{0.33@50^\circ}{0.33@-130^\circ} = 0.67@180^\circ$$

$$\text{PrefaultIC}_{\text{alpha}} = \frac{I_{RLY-2}}{I_{RLY-1}} = \frac{0.33@-70^\circ}{0.33@110^\circ} = 0.67@180^\circ$$

$$\text{PrefaultI0}_{\text{alpha}} = \frac{I_{RLY-2}}{I_{RLY-1}} = \frac{0.00@0^\circ}{0.00@0^\circ} = 0.00@0^\circ$$

$$\text{PrefaultI2}_{\text{alpha}} = \frac{I_{RLY-2}}{I_{RLY-1}} = \frac{0.00@0^\circ}{0.00@0^\circ} = 0.00@0^\circ$$

$Fault1IA_{alpha} = \frac{I_{RLY-2}}{I_{RLY-1}} = \frac{0.04@95^{\circ}}{0.06@-88^{\circ}} = 0.67@183^{\circ}$	$Fault2IA_{alpha} = \frac{I_{RLY-2}}{I_{RLY-1}} = \frac{0.04@95^{\circ}}{0.06@-85^{\circ}} = 0.67@180^{\circ}$
$Fault1IB_{alpha} = \frac{I_{RLY-2}}{I_{RLY-1}} = \frac{3.10@-171^{\circ}}{4.65@9^{\circ}} = 0.67@180^{\circ}$	$Fault2IB_{alpha} = \frac{I_{RLY-2}}{I_{RLY-1}} = \frac{2.44@9^{\circ}}{3.66@-171^{\circ}} = 0.67@180^{\circ}$
$Fault1IC_{alpha} = \frac{I_{RLY-2}}{I_{RLY-1}} = \frac{3.10@8^{\circ}}{4.65@-172^{\circ}} = 0.67@180^{\circ}$	$Fault2IC_{alpha} = \frac{I_{RLY-2}}{I_{RLY-1}} = \frac{2.44@-170^{\circ}}{3.66@-10^{\circ}} = 0.67@180^{\circ}$
$Fault1I0_{alpha} = \frac{I_{RLY-2}}{I_{RLY-1}} = \frac{0.005@-71.73^{\circ}}{0.007@108.27^{\circ}} = 0.714@180^{\circ}$	$Fault2I0_{alpha} = \frac{I_{RLY-2}}{I_{RLY-1}} = \frac{0.001@-31.30^{\circ}}{0.002@148.70^{\circ}} = 0.5@180^{\circ}$
$Fault1I2_{alpha} = \frac{I_{RLY-2}}{I_{RLY-1}} = \frac{1.812@98.47^{\circ}}{2.718@-81.53^{\circ}} = 0.67@180^{\circ}$	$Fault2I2_{alpha} = \frac{I_{RLY-2}}{I_{RLY-1}} = \frac{1.388@-80.46^{\circ}}{2.082@99.54^{\circ}} = 0.67@180^{\circ}$

Figure 3-36: Test Case #3 Alpha Plane Calculations

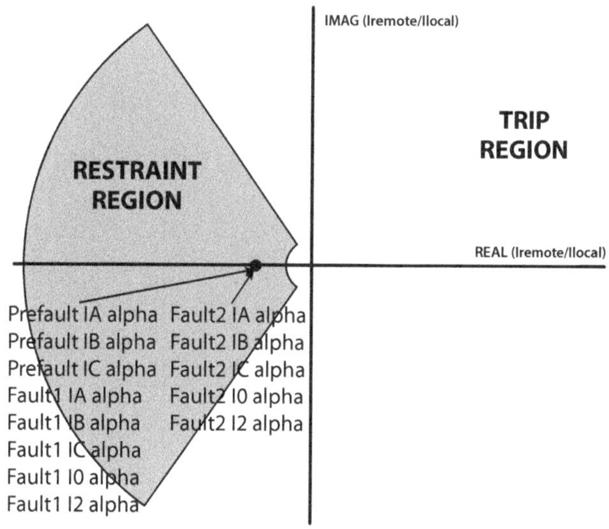

Figure 3-37: Test Case #3 Alpha Plane Graphed Calculations

C) Line Differential Test Results

While it is possible to test the slope or alpha plane characteristics, end-to-end testing is designed to apply real faults and observe reactions. With this in mind, the simple differential philosophy of "only trip when fault is between the two relays" still applies with the new characteristic and the following results will occur when the standard battery of tests are applied.

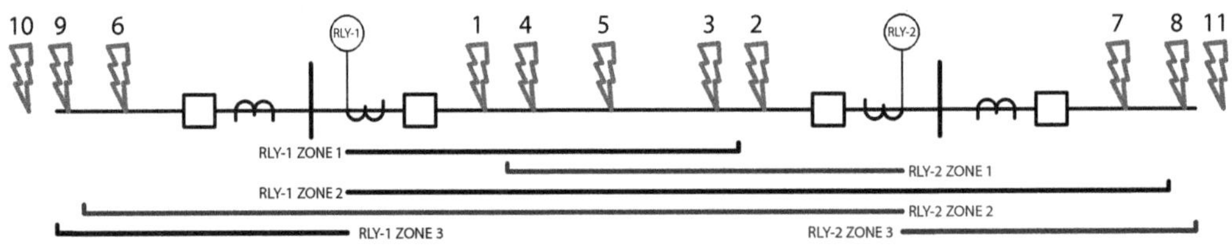

Figure 3-38: End-to-End Test Simulations

TEST	RLY-1	RLY-2
1	Trip 87 in <3 cycles	Trip 87 in <3 cycles
2	Trip 87 in <3 cycles	Trip 87 in <3 cycles
3	Trip 87 in <3 cycles	Trip 87 in <3 cycles
4	Trip 87 in <3 cycles	Trip 87 in <3 cycles
5	Trip 87 in <3 cycles	Trip 87 in <3 cycles
6	No trip	No trip
7	No trip	No trip
8	No trip	No trip
9	No trip	No trip
10	No trip	No trip
11	No trip	No trip

Figure 3-39: End-to-End Test Results with Line Differential Scheme Applied

Chapter 4

Conclusion

End-to-end testing requires additional software to create test plans, a careful review of the test plans, modern test equipment, and good communication between the relay testers at each end. After all of these items have been resolved, end-to-end testing only requires a few extra test procedures when all the protective relays, communication equipment, and relay settings are working properly.

When you review the test results in the field, these two principles will apply to most communication schemes:

- Faults inside the zone of protection (current at the same angle on both ends) will be isolated more quickly by both relays than they would without the communication scheme.
- Faults outside the zone of protection (currents at each end 180° apart) will be ignored or have significant time delays.

The details of the different communication schemes (POTT, PUTT, etc.) can be confusing, but all the schemes provide the same results as per the following chart.

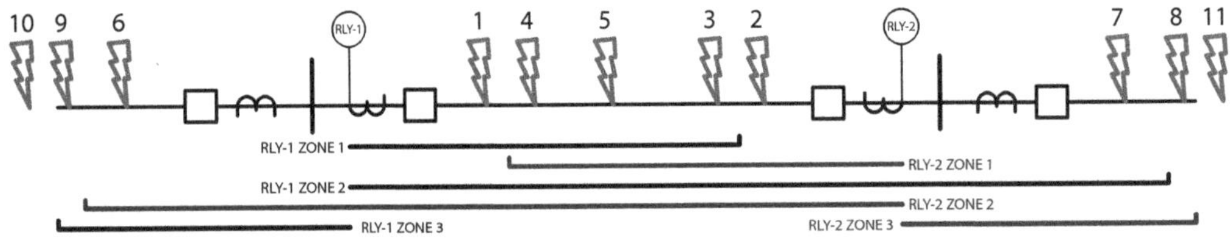

Figure 4-1: End-to-End Test Simulations

	POTT		PUTT		DCB		Pilot Wire		Phase Comparison		Line Differential	
	RLY-1	RLY-2	RLY-1	RLY-2	RLY-1	RLY-2	RLY-1	RLY-2	RLY-1	RLY-2	RLY-1	RLY-2
1	Trip Zone-1 in <3 cycles	Trip Zone-2 in <6 cycles	Trip Zone-1 in <3 cycles	Trip Zone-2 in <6 cycles	Trip Zone-1 in <3 cycles	Trip Zone-2 in <6 cycles	Trip 87 in <3 cycles	Trip 87 in <3 cycles	Trip 87 in <3 cycles	Trip 87 in <3 cycles	Trip 87 in <3 cycles	Trip 87 in <3 cycles
2	Trip Zone-2 in <6 cycles	Trip Zone-1 in <3 cycles	Trip Zone-2 in < 6 cycles	Trip Zone-1 in <3 cycles	Trip Zone-2 in <6 cycles	Trip Zone-1 in <3 cycles	Trip 87 in <3 cycles	Trip 87 in <3 cycles	Trip 87 in <3 cycles	Trip 87 in <3 cycles	Trip 87 in <3 cycles	Trip 87 in <3 cycles
3	Trip Zone-1 in <3 cycles	Trip Zone-1 in <3 cycles	Trip Zone-1 in <3 cycles	Trip Zone-1 in <3 cycles	Trip Zone-1 in <3 cycles	Trip Zone-1 in <3 cycles	Trip 87 in <3 cycles	Trip 87 in <3 cycles	Trip 87 in <3 cycles	Trip 87 in <3 cycles	Trip 87 in <3 cycles	Trip 87 in <3 cycles
4	Trip Zone-1 in <3 cycles	Trip Zone-1 in <3 cycles	Trip Zone-1 in <3 cycles	Trip Zone-1 in <3 cycles	Trip Zone-1 in <3 cycles	Trip Zone-1 in <3 cycles	Trip 87 in <3 cycles	Trip 87 in <3 cycles	Trip 87 in <3 cycles	Trip 87 in <3 cycles	Trip 87 in <3 cycles	Trip 87 in <3 cycles
5	Trip Zone-1 in <3 cycles	Trip Zone-1 in <3 cycles	Trip Zone-1 in <3 cycles	Trip Zone-1 in <3 cycles	Trip Zone-1 in <3 cycles	Trip Zone-1 in <3 cycles	Trip 87 in <3 cycles	Trip 87 in <3 cycles	Trip 87 in <3 cycles	Trip 87 in <3 cycles	Trip 87 in <3 cycles	Trip 87 in <3 cycles
6	Trip Zone-3 in 60 cycles	Trip Zone-2 in 20 cycles	Trip Zone-3 in 60 cycles	Trip Zone-2 in 20 cycles	Trip Zone-3 in 60 cycles	Trip Zone-2 in 20 cycles	No trip	No trip	No trip	No trip	No trip	No trip
7	Trip Zone-2 in 20 cycles	Trip Zone-3 in 60 cycles	Trip Zone-2 in 20 cycles	Trip Zone-3 in 60 cycles	Trip Zone-2 in 20 cycles	Trip Zone-3 in 60 cycles	No trip	No trip	No trip	No trip	No trip	No trip
8	No trip	Trip Zone-3 in 60 cycles	No trip	Trip Zone-3 in 60 cycles	No trip	Trip Zone-3 in 60 cycles	No trip	No trip	No trip	No trip	No trip	No trip
9	Trip Zone-3 in 60 cycles	No trip	Trip Zone-3 in 60 cycles	No trip	Trip Zone-3 in 60 cycles	No trip	No trip	No trip	No trip	No trip	No trip	No trip
10	No trip	No trip	No trip	No trip	No trip	No trip	No trip	No trip	No trip	No trip	No trip	No trip
11	No trip	No trip	No trip	No trip	No trip	No trip	No trip	No trip	No trip	No trip	No trip	No trip

Figure 4-2: End-to-End Test Results with All Schemes

You can make the test procedure even more comfortable by practicing before the test date:

1. Set up two test-sets on opposite ends of a table so that you are not able to see each other's actions to simulate being at remote locations.
2. Connect A phase voltage and current from both test-sets into any phases of the relay.
3. Load the identical file into both test-sets. (You would load the same test case but different files when performing the test for real)
4. Perform the test procedure as you would in the field.
5. Review the event record waveform from the relay. The voltages and currents should be identical.
6. Repeat until you feel comfortable.

Bibliography

Benmouyal, Gabriel; Mooney, Joe B.; *Advanced Sequence Elements for Line Current Differential Protection*
Schweitzer Engineering Laboratories
Pullman, WA, www.selinc.com

Roberts, Jeff; Tziouvaras, Demetrios; Benmouyal, Gabriel; Altuve, Hector J.; *The Effect of Multiprinciple Line Protection on Dependability and Security*
Schweitzer Engineering Laboratories
Pullman, WA, www.selinc.com

Ariza, J.; Ibarra, G.; *Application Case Of The End-To-End Relay Testing Using GPS-Synchronized Secondary Injection in Communication Based Protection Schemes*
Megger, U.S.A.
CFE, Mexico

Araujo, Chris; Horvath, Fred; Mack, Jim; A *Comparison of Line Relay System Testing Methods*
National Grid Co.
FPL Seabrook Station
Engineering Laboratories, Inc.
20060925 • TP6251-01

Manta Test Systems; *Time Synchronized End-to-End Testing of Transmission & Distribution Line Protections with the MTS-5000*
Application Note: AN506
Manta Test Systems Inc, www.mantatest.com

Schweitzer Engineering Laboratories; *Applying the SEL-321 Relay to Directional Comparison Blocking Schemes*
SEL Application Guide
Pullman, WA, www.selinc.com

Schreiner, Zeljko; Kutner, Reinhard; *Remote Controlled Testing of Communication Schemes for Power System Protection Using Satellite (GPS) Synchronization and Modern Communication technology: A New Approach*
Omicron Electronics GMBH, Austria

WSCC Telecommunications and Relay Work Groups; *Communications Systems Performance Guide for Protective Relaying Applications*
Nov 21, 2001

Guzman, Armando; Roberts, Jeff; Zimmermann, Karl; *Applying the SEL-321 Relay to Permissive Overreaching Transfer Trip (POTT) Schemes*
SEL Application Guide
Pullman, WA, www.selinc.com

Mooney, Joe; *Communication Assisted Protection Schemes*
SEL Application Guide
Schweitzer Engineering Laboratories; Hands-on Relay School
Pullman, WA, www.selinc.com

Tang, Kenneth; *Dynamic State & Other Advanced Testing Methods for Protection Relays Address Changing Industry Needs*
Manta Test Systems Inc, www.mantatest.com

Tang, Kenneth; *A True Understanding of R-X Diagrams and Impedance Relay Characteristics*
Manta Test Systems Inc, www.mantatest.com

Blackburn, J. Lewis (October 17, 1997); *Protective Relaying: Principles and Application*
New York. Marcel Dekker, Inc.

Elmore, Walter A. (September 9, 2003); *Protective Relaying: Theory and Applications, Second Edition*
New York. Marcel Dekker, Inc.

Elmore, Walter A. (Editor) (1994); *Protective Relaying Theory and Applications (Red Book)*
ABB

GEC Alstom (Reprint March 1995); *Protective Relays Application Guide (Blue Book), Third Edition*
GEC Alstom T&D

Schweitzer Engineering Laboratories (20010625); *SEL-311C Protection and Automation System Instruction Manual*
Pullman, WA, www.selinc.com
Markham, Ontario, Canada, www.geindustrial.com

GE Power Management (1601-0089-P2 (GEK-113317A)); *D60 Line Distance Relay: Instruction Manual*
Markham, Ontario, Canada, www.geindustrial.com

Young, Mike and Closson, James; *Commissioning Numerical Relays*
Basler Electric Company, www.baslerelectric.com

Avo International (Bulletin-1 FMS 7/99); *Type FMS Semiflush-Mounted Test Switches*

VALENCE
ELECTRICAL TRAINING SERVICES

https://relaytraining.com store@relaytraining.com 303-250-8257

The Relay Testing Handbook
Principles and Practice
Chris Werstiuk